Africa Is That Story That We Tell Ourselves

FRANCIS OTOLO

Copyright © 2022 FRANCIS OTOLO

All rights reserved. No part of this book may be reproduced, or stored in a retrieval system, or transmitted in any form or by any means, electronic, mechanical, photocopying, recording, or otherwise, without express written permission of the publisher.

ISBN: :9798846426160

Cover design by: Vadtecks

Printed in the United States of America.

DEDICATION

To love you by the writings, the covenant of the Ark, and the exodus of the heart, and to the several mist.

To the creators, and inventors, and their aspirants.

To the saints, and the prophets, and of all that were slain upon the earth, and the martyrs of Jesus earth,

CONTENTS

	Dedication	iii
	Table of Contents	v
	Forward	vii
	Preface	xi
	Acknowledgement	xv
	Introduction	xvi
1	Prayer	1
2	Be Content	7
3	Problem	13
4	Learn by observing	18
5	Understanding and Wisdom	24
6	Courage	29
7	Live Culture	34
8	Be a day dreamer	38
9	Ever Living Faith	43
10	Sorrow in the Way	47
11	Joy in the way	53
12	Be Focused	57
13	Measuring	61
14	Love and Charity	66
15	An Encounter with the World	73
16	This book and other experiences	78

FORWARD

This may be the only book right now in the world that can teach and help you practice creative wisdom to the very limits directly from the bible, and, by the way that leads to creating, making and forming things. I can confidently show you that the bible is. And that you can achieve it in your life. Because this knowledge is rare as I have seen but of the LORD divine mercy it is given to me. And take genuine practice, with it do something you have not done before, soar beyond the level of principles, and then you will find good success there. Now this is chiefly written from personal observation, and experience; since some areas may apply to you differently when exposed to similar conditions. Because of your human nature, and that there are two types of human that exist – alien and non alien.

I have brought this book to you under serious direct threats to my life from people who don't want these secrets revealed to you. Some said it is the way of the black monks to draw near an end. And it is written under different types of suppressive conditions even to my physical health, either to cause delay or to limit what is included and reduce the content quality. This work is a native and direct work – Christ is the fountain head. I confirmed these things from my experience in the siege mentioned in this book. And the bible is the basis of this journey, hence its references everywhere. And that all references outside the bible were to help me push forth the light, so that you may be confident and trust in the work before you.

I consider Baxter's **'A Request To The Rich'** from his Work: **The Poor Man's Family Book** as a more fitting text to do the forward here, because they were written at a time when the knowledge of what I am going to talk about was also widely revealed, and known, and practice among the people between 1300 and 1900. And record also showed that this led to the peak of transformation (or culmination) at the end of the nineteenth century. The same period the bible received one of its best translations – the KJV. As a light to that generation, the highest number of inventions took place within the period till date, when compared to earlier centuries, as some inventions were started and discovered within that period.

If you truly know Christ, science will be at your mercy. Now the substance of science is in Christ, and when it is understood, in the path of excellence, then can it be widely spread in the schools. Tell me that which is not modeled from the things of God among current technology and innovations? From your gun, cars, phones, computer, internet they were all inspired, and mimicry, and copied from creation and heavenly things. But the greatest glory is in God's unit of being simple.

The Forward: A request to the rich

"THIS book was intended for the use of poor families, which have neither money to buy many, nor time to read them: I much desired therefore to have made it shorter; but I could not do it, without leaving out that which I think they cannot well spare. That which is spoken accurately, and in few words, the ignorant understand not: and that which is large, they have neither money, leisure, nor memory to make their own. Being unavoidably in this strait, the first remedy lieth in your hands: I humbly propose it to you for the souls of men, and the comfort of your own and the common good, on the behalf of Christ, the Saviour of your souls and theirs, that you will bestow one book (either this or some fitter) upon as many poor families as you well can. If every landlord would give one to every poor tenant that he hath, once in his life, out of one year's rent, it would be no great charge in comparison of the benefit which may be hoped for, and in comparison of what prodigality consumeth. The price of one ordinary dish of meat will buy a book: and to abate, for every tenant, but one dish in your lives, is no great self-denial. If you, indeed, lay out all that you have better, I have done. If not, grudge not this little to the poor, and to yourselves; it will be more comfortable to your review, when the reckoning cometh, than that which is spent on pomp and ceremony, and superfluities, and fleshly pleasures. And if landlords (whose power with their tenants is usually great) would also require them seriously to read it, (at least on the Lord's days,) it may further the success. And I hope rich citizens, and ladies, and rich women, who cannot themselves go talk to poor families, will send them such a messenger as this, or some fitter book, to instruct them, seeing no preacher can be got at so cheap a rate. The Father of spirits, and the Redeemer of souls,

persuade and assist us all to work while it is day, and serve his love and grace for our own and other men's salvation. Amen.
Your humble Monitor, Aug. 26, 1672. RICHARD BAXTER.

If I cannot buy many books, then give me one which is rich. Therefore not a threat to the soul but little companion; how that will make suit, and lead you on to God. I have written down some truths and statements boldly as you can see in this forward and over all pages of this book, and it appears the proof are not here and afar off. However, they all lie hidden in this book. The biggest mysteries of the world are not afar off but within. And that the overcoming and triumphing of it, is in how you enter in to it. Here in is a sure guide and a code written partly in blue.

How Creativity leads to invention or true novelty (Gen 1-5).
In creative people are all the elements formed and made that prospers other abilities. Africa is that story we tell ourselves. And that story is our source of invention or destruction; and to make or mar us.
Now take a look at the number of novelty works and patents out there. You can do this research over the internet which is just a finger tip away for many.
If you are satisfied completely with the work from this continent then this book is not for you.

How well we are going to progress depends on how much of the underlying truths behind all novelty work out there that we are exposed to. The more of such truths we are exposed to the greater the chances of survival for this generation or next. If we can get it right, then the coming generation would be more glorious than this one. Use the little space on top of the next page to make a commitment to be part of a people that set a foot print for others to be more glorious.

AFRICA IS THAT STORY THAT WE TELL OURSELVES

The foundation or source of this glorious work that the west and the rest of the world got graciously in the past is also very secretly hidden and unknown among majority of them. Many enter into it again by luck.

The way to this source or foundation is the greatest desire of all nations. Many new patents are built from or on previous existing principles and truths.

In the bible, the priest and the prophet were the people consulted on matters of invention. Specially, the prophet and the priest had different unique roles. So that wherever the house of God is, there is solution to all things.

However, the priest and prophet functions are also present and well represented differently in the western or other more prosperous continents from what is done here in many countries in Africa. Now the priestly or prophetic occupation everywhere is trading out different kinds of products because the interest is different.

For the pastors, if God truly called you as you say, then find Him, and you will know these things. Why the whole Africa is nowhere or barely found in patents application records

PREFACE

Africa is that story that you tell yourself. Is not this the thing you do, your confidence, the hope of your generation, and the creativeness of your ways? What good shall your life do you after this story? Certainly, even as I have known, they that invest in wisdom, and serve God, and do service to man, reap the same.

What is Africa's biggest problem?
Is it not able to pioneer meaningful visions and change apace with other nations with whom they started, among other numberless shortcomings? As a continent that is wise may be profitable to herself?

What is the solution?
Now the solution is not in changing the narrative. Because we have change it countless times, and tried to get it before our founding fathers, and nothing has changed. But in knowing, and, owning the narrative itself. To do this, ***"And ye shall measure from without the city on the east side two thousand cubits, and on the south side two thousand cubits, and on the west side two thousand cubits, and on the north side two thousand cubits and the city shall be in the midst"(Num 35:5).*** That is we have to take, see, know, and own it, which leads us to creating, making and forming things. Which tend to where it leads, and its nature and richness, and fruitfulness and finally security and safe landing? Though stopped and filled with earth, the narrative is simple. And that the purpose of this book is to help you rediscover it and own it. And to discover it again, is to start living a meaningful life. When it is righted, this life we intensify to empower the strength as well as the weakness by way of besieging it.

The Prophets and Inventions:

The portion of Jacob is not like them:
for he is the former of all things;
and Israel is the rod of his inheritance:
The Lord of hosts is his name.

and to weave all things, and to invent all new things.

It is God's privilege to conceal things,
and the king's privilege to discover and invent.

And the harp, and the viol, the tabret,
and pipe, and wine, are in their feasts:
but they regard not the work of the Lord ,
neither consider the operation of his hands.

Therefore my people are gone into captivity,
because they have no knowledge:
and their honourable men are famished,
and their multitude dried up with thirst.

That chant to the sound of the viol,
and invent to themselves instruments of musick, like David;

This also cometh forth from the Lord of hosts,
which is wonderful in counsel,
and excellent in working.

and they shall come, and see my glory.

For you or the African youth

Our fathers were genius and got it, now are we losing it? For man or the African youth, to make this happen. We have few great schools, few right people in government, and few great companies and some very few who live abundant lives principally because they understood and own the narrative. And that truly successful people, have always

AFRICA IS THAT STORY THAT WE TELL OURSELVES

sought a way to improve their lives, and with all their heart, body and soul in accordance with God's laws. They could not accept a second hand God. And countless of such people were determined and remain steadfast to seek their God but not men's commandment and lies.

All things: The application of the Word

The words we speak in all spheres of life are from the Word of God, thus from God, and as powerful as God Himself. Not only to be taken from theist stand point. And that though some are atheist, but are chiefly ignorant of this fact, that the Word is their life. Also many Christians are trap in not knowing this and are helplessly deficient in restoring ancient paths and not able to sail into wider civilizations from thence and making things happen within. Now, the knowledge of God's relationship with man is the starting point of all things. And that many Africans don't know their prophets. Still some kill their prophet. But the prophet is the only true pathway into the universe. For the atheist, this is not to be viewed only for some as a religious book but to affirm again that God is the actual source of solution to the great and small problems in the world. Many people seem to short change themselves because of their religious belief – a fruitless alibi. Because that they claim to know God, but know Him not.

In the church, we pray about original creative schools, (try to bring up students who can invent and create), and companies (looking for highly skilled and creative employees and inventors) and training (develop/create) or education centres, people pray about other abilities in this area and needs. Many companies have little or nothing to show for the effort. However, this kind of life puts you ahead of others and need to be secretly and deliberately and independently develop to help humanity and solve societal problems. And that our limitations are limitations because we call them limitations. Because we have not in contentment sufficiently acquainted ourselves with the primary technical know-how knowledge in practice or independently ride the bicycle without an aid. While we should fall and rise again, and again, we should also independently put crafts in the air, and allow them to soar or fall and be destroyed, until reach the place of success. And we must jump off

the cliff without considering where we are landing if we are so bidden. Now this book is meant to make you, and to give you right understanding that can develop your ability to genuinely use the Word of God as a true source to practically find success, and that it can get you started immediately in any endeavor, and especially where high creativity is required. When you understand that the Word is the only true source to all things, then your winning edge is set.

No family library should be without this book. To develop an effective economic resilient people, where will you look into? Until we come to our ethnic (Jesus) places that ancient fountains Jesus built, we cannot get it right quickly with the rest of the world. Even, the hidden springs of the church or religion. The root is the same; the true source is one. Whether you are a Muslim, Christian, traditionalist or whatever you call yourself. Let us spring forth from the original fountain head, which is our foundation and Christ Himself.

'And Jesus answered and said unto him, What wilt thou that I should do unto thee?'(Mark 10:51)

What is your desire? If I am clear with it and my eyes is single, then the whole world help to devise my success. And will do it for you more perfectly than conventional wisdom has inspired.

When we have asked God for help in prayers, for some or the majority of the people there remains a great veil between God, and the solution and man, which takes undying love for God and belief, and courage to step in and to be supersensitive while observing to notice. Not everyone will hear a loud voice in the garden to acknowledge the presence of God, but for some only few it would be very loud, and others no voice and neither was it observed. To the successful observers they have many mothers and fathers who is their mothers and fathers? How do we partake in this mystery?

However, I have written this book to show people how to secretly develop themselves on the area and practically become creative and solve daily challenges with it. The chapters are meant to open you up so that the word may take effect and you interact with it

ACKNOWLEDGMENTS

To everyone on my publishing team, I give thanks. And special thanks to Monday Eto and the management of Shammah Christian Center for the invaluable help given to me in their training in the course of publishing this book

INTRODUCTION

She said, "Long ago people used to say, 'If you want good advice, go to the town of Abel to get it.' The answers they got here were all that was needed to settle any problem. 2 Sam 20:18 CEV
Return to the stronghold [of security and prosperity], Zech 9:12

And that you are about to travel into the land of uses. Let me take you through the perfect road, the perfect road is calm and easy. Here in this book is a journey to reveal secrets from the fifteen chapters on how to daily tackle problems in the original way with God. Our focus is ever-increasing the ability to create, make and form things, using the bible, as primary source of guidance.

I am as one writing from the trench with working coverall stained and dirty. And taking responsibility or rejecting complaints in things not properly done and rewards for tasks successfully completed. Now this book is put together by creative elements that are practically put to use. And elements that enable you to achieve the plan of this book have been successfully tested and are being use. Here discouraging challenges and problems are 'live', when you are embarking on the journey. Every day they make story of you. I call them live because they pop up unexpectedly and some are there to stop you, while others may be there to help you.

In the market today the priciest commodity is principal wisdom and knowledge of how things work in the world. They are highly sought after who can create but not found. And that you are God's source of creativity and wisdom, and is the solution and pathway to national development, which is creation of employment and helping of our common household and poverty eradication. The reason many people are not creative and solution are covered in this work. Also, the ancient building blocks from the scriptures for creative life are hereby captured. Creativity is not meant for the only few that found the way, it is for anyone that can also learn and follow the path. Many children are going to graduate the same graduates that

AFRICA IS THAT STORY THAT WE TELL OURSELVES

their parents were when they graduated. Or I could as well finish this book the same book as others poorly done without the real substance. There is an otherworldly creative space, the garden or some Island – a place for the Einstein's of this age, where our ancient treaded. This work is not to show you the place but to bring you where you can find the place. "But where shall wisdom be found? and where is the place of understanding? And that it is the most profound source of knowledge. If you truly understand and do the things you will learn here, you will find the ancient city, the pleasant land, or what some biblical translations called the land of desire. It is owned by God alone, and it is the desire of all nations. In the land there is access to all creative work and all nations is not present. It is not available in the knowledge market. When you come to the place of all inventions, innovations and creative work you will find no competitor there.

Today, the newspaper pages tell of lost jobs and redundancy in secured corporations across the globe. As a result many are afraid but there is a fountain of all solutions. You can work and independently create jobs for others and yourself as Christian using your faith and the bible. The words of the LORD Jesus are as real as in quantifiable faith. Christ is also in real data. Christ is measurable. Joseph numbered Egypt. Faith is a substance, not an invisible thing.- it is visible. With the knowledge of God you can do all sector creativity but the deeper things may not be available to you, unless you specialize in that area. Now we can all learn how to become solution providers in our society. From principal wisdom that can help us solve big problems and needs. And to undertake meaningful researches and ventures. Let the youth and anyone interested in wisdom read this book. With this book if you cannot find the land, you will still do great things. For those who overcame, and on the minimum, God's final arrival plan reads, 'And for his diet, there was a continual diet given him of the king of Babylon, every day a portion until the day of his death, all the days of his life.(Jer 52:34). One of the best gifts you can ever give to the poor is this kind of books.

PRAYER

Therefore hath thy servant found in his heart to pray this prayer unto thee.-2 Sam 7:27
Yea, thou castest off fear, and restrainest prayer before God.-Job 15:4KJV

Because many think that prayer is meant for some people, and that it should be left for religious people, and pastors, that many also in our generation lost how to pray, and their prayers. Unless, we know how to pray, we cannot know how to deploy prayer as an effective mercy tool in our daily lives. When you pray, pray an ever living prayers. Howbeit, the most powerful inventors and wise people in the world are those who can, and know how to pray – show me the best scientist, many have used the most powerful secrets of ever living prayers. Because the blessing is in obeying the commandments, they in some ways met this requirement, and obtained outstanding results. For some I can show you what they did from the scripture. For example when Einstein said imagination is more important than knowledge, he simply told you about applied meditation. And many other such things Einstein did which are outright secret of prayers. If this is true, then there are either secrets (hidden) on prayers or many don't know how to do it. However, I have come with some ever living ways to pray. For many a psalm, and a hymn. Others use their own methods and it work effectively at any time for them.

Not for everybody but what works for all.

Now the most important question is, does your method or means work always? The word of God on prayer in the bible is truth, and if it does not work for us, it may be because we don't know how to pray or we did not understand the bible. When Jesus said, if ye ask anything in my name, I will do it. He did not place a condition, if only we should ask in his name. Many people will not know that the cost of prayer is sometimes persistence. When your prayers are not answered immediately, then you have to continue until you have answers. Until you learn and understand how to pray through this method, don't stop. By this one can come to a place of ever living prayers.

Early travail before going out of Egypt

'For God giveth to a man that is good in his sight wisdom, and knowledge, and joy: but to the sinner he giveth travail, to gather and to heap up, that he may give to him that is good before God. This also is vanity and vexation of spirit.'(Eccl 2:26)

From this stage of travail you should move to a place of automatic key, from where you had used manual methods. If you do well angels will teach you by revelations, and show you secrets on what works and how to pray effectively, using the bible, hymns and psalms. After you have learnt how to pray, you may have to deal with prayer robbers, if they are in your family. Then you should get a chance to know dream and the accuser of brethren because they make sure you are denied access. This very fact Isaiah wrote: *'Which justify the wicked for reward, and take away the righteousness of the righteous from him!(Isa 5:23)*

I have seen this happen to me again and again. My righteousness was taken away, when I pray, they diverted the good things elsewhere.

This is one of the chief roles of the Egyptians in the bible, Moses writes: *"And Pharaoh charged all his people, saying, Every son that is born ye shall cast into the river, and every daughter ye shall save alive.'* Now the good things are represented also as daughter. While all things of knowledge is referred to as son. There are many dedicated Christians that are passing through this experience. The enemy abuses all the persistent effort. To those that know their God their adversaries are judges because I have also seen that the enemy is one of the best source of knowledge or the one who show you the way; be careful, love and charity and forgiveness should be your banner. For those that know their God their enemies are judges *'For their rock is not as our Rock, even our enemies themselves being judges.'(Deut 32:31).* When last did you drink from the old spiritual Rock? The scripture also says he who believes out of his heart shall flow rivers of living waters. Your Rock is in your heart in God's word. Today, in anger many know not the Rock but have made it a target of destruction.

When God give you the mystery or wisdom of how most of our prayers work, you will see and learn how it is done. How shall we escape? I have also given or show you how you can be delivered from it. You should serve God to the point of sensing and discovering this act. And that prayer is may be also counted or measured for righteousness. In our everyday language, many people use some foolish words to describe very powerful methods of prayer such as *"lie lie mumu"*. That I got from an easterner, an Arabian and it is commonly used among the people. I don't know how we got it but it is richer than you can ever imagine if you know it. These among others are proof to show you that there is more than meet the eyes on prayer, in the secret things of God. To the world is foolishness but to God's children is a wonder. Even these descriptions are not figures of speech or proverbs. Know for certainty that understanding have they who know how to pray, even as it is difficult to find a man, so is it with prayerful men. In all history and especially in the bible, there have been search for men(a man). Many people don't even know what a man is. Get a man then you will have a teacher of prayer. When you know how to pray; others come to you in prayer. This particular mystery is broad and simple but it took me years to understand. It is of the LORD's divine mercy that those who seal, and unseal showed it to me. And they could be genius in your field or in area of your needs. All these things are written in the bible and are well hidden in many instances, so that one cannot learn what is going on in the spiritual world. They are invisible but more than visible; as a matter of fact very visible.

But I found none
For some, in the life of a man, the woman is connected to the man. This is not closing and opening of eyes, at that connection is live and powered by the LORD, and you do more than feel it, you touch the experience, and here you touch the spirit, not consciously and feeling but by a more mechanical and electrical mystery. No matter the distance apart from the wife, for some, even me, and I can hear, and talk to me any time. All these things are written all over the bible from Geneses to Revelation. This is a death knell that destroyed or demystified the lies about the Holy Spirit and spirits (But next time be very careful what you do in the secret, the birds will carry it away) who are involve in all secret things of your life. And I can feel my soul, and it carries my body. And it sensate anything, like everything of all the senses; this ability comes from the soul. When others crings and plot, against me, and about my soul, and I hear them, not when I am sleeping but awake. And the spirit can be under your feet in your body without getting hurt. You can feel it.

To be very prayerful, you can use other forms which are everlasting, and are found in the scripture; I know that in the fourth chapters of Deuteronomy, and also in Daniel is it found. I am not writing to explain, but to let you know that there are fountains deep in the things of God. Therefore enter places in the word, or learn it, and search for free things of grace. How you take more, and try some presumptuously – gather to you living springs.

For scriptures, in patience attentively with, and look.
Learn to spend time understanding visions and objects in revelations you received, even if it takes years to unravel them — keep them in mind always. Has that anything to do with the fear of God or called the fear of the LORD? But try to know, if it is worth seeking out. Then to pay any attention to revelations- prepare and work for it. When praying, be fervent and listen. Learn more by being supersensitive to expectations in any form. Also you can achieve this by observing outcomes and statutes, and testimonies of people, and be attentive to secrets, especially from renowned men of God and scripture. If you are able to find any, it will be permanent, and be useful to you and others with you.

Answers
Now answers to prayers can come from anywhere from pulpit or any preacher where you instantly made the prayer or a day before, from passerby speaking to his friend or even your little child talking outside. And that answers to your prayers come from countless sources. Consider a mathematically quantifiable thing to prayer. One plus one is equal to two. Learn to do well and surely it will work most times for you like that. Now most Christians speak, of laying up treasures; or bank account in heaven, should we not also gain access to it any time? If I cannot access it at any time then something is wrong with my faith. Then I have no treasures in heaven, or I am not sure. I simply have no faith. Faith is in the name of a man. Yea, a child of God. According to the book of Hebrew, 'now faith is a substance of a thing hope for, evidence of a thing not seen. By prayer or an open cheque we receive that substance because it was already laid up by our Father in heaven. The reason man is very difficult to find is because we don't know that faith is in the name of a man, even in the name of Jesus Christ. Like the heaven and the earth, the life of a man is in correspondence, and can be spiritually destroyed by the man's prayer, and that may also affect others. These are ancient

paths and are very present to any man. Your life spiritual progress and activities can be reset. Like, you do to a phone. These things, you will see not a ghost science. In a moment of minutes not hours.

Consider your ways
and the Lord granted his prayer, and Rebekah his wife became pregnant.- Gen 25:21

Continue in prayer, and watch in the same with thanksgiving; (Col 4:2)

And the prayer of faith shall save the sick, and the Lord shall raise him up; and if he have committed sins, they shall be forgiven him.(James 5:15)

Woe unto you, scribes and Pharisees, hypocrites! for ye devour widows' houses, and for a pretence make long prayer: therefore ye shall receive the greater damnation. (Matt 23:14)

When my soul fainted within me I remembered the Lord : and my prayer came in unto thee, into thine holy temple.(Jonah 2:7)

Also when I cry and shout, he shutteth out my prayer.(Lam 3:8)

O thou that hearest prayer, unto thee shall all flesh come.(Ps 65:2)

27 Thou shalt make thy prayer unto him, and he shall hear thee, and thou shalt pay thy vows.

28 Thou shalt also decree a thing, and it shall be established unto thee: and the light shall shine upon thy ways.
(Job 22:27-28)KJV.

BE CONTENT

Now therefore be content, look upon me;
for it is evident unto you if I lie.- Job 6:28

Do violence to no man, neither accuse any falsely;
and be content with your wages.-Luke 3:14

These stories are deep parables and arcana of heaven not as you seem to have concluded on them. To be content is to sit still, place you, and find you perfectly fit within God's original plan. If you be content as a person, others will challenge you. Then be ready for wars, from those who are not. If you did not experience this phenomenon you have not come to a place of contentment. To do this, knowledge is required and you may have to fight and win wars. The wars bring you to the place of contentment. To be content is having sufficient knowledge to be self-reliant in a siege. Now the siege is the vehicle into your future. To be content is to live by, or put trust in knowledge acquired; get a masterful use of them. If you are not content, you will not be patient enough to see the miraculous that the knowledge contains. Until you come to such a siege, people will continue to direct your life. You will not come to see the brooks, the floods, and the honey and………. This is where people are rejected and abandoned. Give a secret recess to your life. If you lose the battle to others you will not come to the place where you can go on personal recess. The Eureka place is a busy place for the brain and your imaginations. When you do this well with God, you will possess your soul, and you will find the ancient Garden of Eden. If I can discover these things, then you also can do it.

Sit solitary
A genuine creative work is not done following the herd. When you sit solitary, solid and native books are written, and, from deliberate isolation with a contented mind. Because you are learning, equip yourself with the right books when living a contented life and read them.

The leading of the Holy Spirit
Like Moses gather all the elders you have ever met in the course of your Christian life and meet Pharaoh King of Egypt. If you don't have any elder to go with you, it means you are not born again. Or gone to church before. Now with the bible, your pastor and all the wonderful preachers out there you should have already gathered and encountered some elders with God's help – they are looking at you now, you don't know? If you don't know or don't have don't worry. The bible is the number one book you should take with you. Pharaoh King of Egypt is also knowledge in any form, no matter how deep or advance. Because prayer is what you need to succeed in the siege. Without prayer you cannot be content. In the siege of contentment, possibilities are also endless that is why prayer is key. This is a state of life that should lead you to any scientific discovery or creative work. And also, build a true spiritual life. At the initial stage, depending on your situation, I suggest you cut off all external help. If you are using this to undertake a complete spiritual journey, pray until you get direct leading of the Holy Spirit – not presumption and mere feelings. If you are fortunate, you should be going on with a burning fire in your hands. Fear not. Avoid contemporary knowledge, when

undertaking spiritual journey in order to make the most of it. And be led by the Holy Spirit in practice faith.

No defeat, Sit Still
Furthermore; you can win countless battles without being defeated in the place of contentment. Here you can comfortably take very high risks without staking your head. Never look for the highest watch tower elsewhere- it is here. When you understand these principles, you will not fail any examination. None can deceive you easily, if you abide in this state well enough. Unless you are content, you can easily be deceived by learned people you trust the most. For Moses, he found strangers; when he was content to dwell with Jethro. Where is the pen of a man? Like Moses or Isaiah, you may write with the pen of a man-*'and write in it with a man's pen.'(Isaiah 8:1).* Consider this, even the covenant promises of God common to any man, if you obey and keep his word. If you go forward you will see these things.

Your Soul and your life
Now many things should also make sense to you concerning your soul. Howbeit, one sit solitary but in an active state. Not passive because of insurrection of hell into the visible world in the period. Then you live from the internal spiritual life consciously and observe the influx into visible. No one can do this without steady revelations. And make amends to both lives. I had once seen myself walking as robot with my very self-watching it. Like a physical object in 3D shape in front of me. That is grace at work. This is the moment, when you perceive activities, even at the level of your soul. 'In patience possess your soul'- Jesus warned us. As we become more thankful for little, and be grateful, shall we undo burdens that keep us from reality? When stones or exceeding blows start coming, in place of drawing water, they distract you but remain undisturbed.
Be patient with everyone, but above all with yourself' –ST. Francis De Sales

Rejoice in the benefits

If the deliberate siege of making you turns spiritual battle, it should be so, and then allow yourself to undergo the full regeneration. Take away the vacuum that would be created by this new life, and use your experience for a creative work. For some the encounter is a complete revelation, and walk with God.

"Greatness it turns out, is largely a matter of conscious choice."

Unless we stop doing many irrelevant things and take away the 'to do list', we can never know how much that can be achieved. When we walk in the way of trusting and relying on Christ's provision; our greatest gift is not only listening but actually drinking of that spiritual drink.

'Come, and let us sell him to the Ishmeelites,
and let not our hand be upon him;
for he is our brother and our flesh.
And his brethren were content.
Then there passed by Midianites merchantmen;
and they drew and lifted up Joseph out of the pit,
and sold Joseph to the Ishmeelites for twenty pieces of silver:
and they brought Joseph into Egypt. Gen 37:27-28 KJV

In the twenty seventh verse, why were his brethren content? If they were not content, then the Midianites would not have passed by. I have not told you, but the merchantmen shared some practical scriptural pieces in Joseph. And that without them(his brethren) being content, Joseph's dreams would not have been fulfilled. Now in Joesph, the Satan or Devil or Lucifer is among or is in his brethren. Where is your Satan? Where it be that he had killed his Satan, what would have happened? I have seen many of the top men of God in our country; their prayer against Satan is in wise and effective judgments. Who is Satan? For some the salt of the siege is Satan or the adversary. Without an oppressor there can never be a siege at the place wherein one needs to be content when tried. No smoke without fire. They (his brethren) wickedly and ignorantly showed him the way, but sadly by the hard and narrow path – he was separated. And God had favored on him. He was also a tree planted by rivers of water. There are secret things in that story not mentioned here. When you are content, and obey God's word in similar conditions and, the wells will spring up. *'And the fear of you and the dread of you shall be upon every beast of the earth, and upon every fowl of the air, upon all that moveth upon the earth, and upon all the fishes of the sea; into your hand are they delivered.' (Gen 9:2).* Now the lesson the LORD gives, who provide your impregnable fortress to life? Never will empty promises or prophesies come your way but real substance as Paul preached in Hebrew. And Joseph dream and his bones which also were thus not made manifest in Shechem. But his brethren sincerely, fearfully and prayerfully repented.

In the same way it is told of many others, and I would site another example from work on Kepler's life:

"Kepler's radical ideas rendered him too untrustworthy for the pulpit. After graduation, he was banished across the country to teach mathematics at a Lutheran seminary in Graz. But he was glad — he saw himself, mind and body, as cut out for scholarship. "I take from my mother my bodily constitution," he would later write, "which is more suited to study than to other kinds of life." While Kepler saw his body as an instrument of scholarship, other bodies around him were being exploited as instruments of superstition.' Johannes Kepler 1630.

Consider your way

Would to God we had been content,
and dwelt on the other side Jordan!- Josh 7:7

And the Levite was content to dwell with the man; - Judg 17:11

And Naaman said, Be content, take two talents…

AFRICA IS THAT STORY THAT WE TELL OURSELVES

And he went out from his presence a leper as white as snow. 2 Kings 5:23-27

And having food and raiment let us be therewith content.(1 Tim 6:8)

And one said, Be content, I pray thee, and go with thy servants. And he answered, I will go. So he went with them. And when they came to Jordan, they cut down wood.(2 Kings 6:3-4)

PROBLEM

"The people come to me and ask me to ask for God's decision for their problem. Ex 18:15-16

*and has made known to me now what we desired of You,
for You have made known to us the solution to the king's problem. Dan 2:23*

"What's your problem with us? Why are you invading our territory?"- Judg 11:12-13

The problem with many people is that they don't have any problem or challenging need. If we have a problem or a real need, then we search for a way to tackle it. Because when we saw the problems and acknowledge them we did nothing to solve them. Therefore, we have no problems to solve. When Hiroshima and Nagasaki was destroyed by the Americans occupation forces during the Second World War, the news reported only these cities but over hundred cities of Japan were flattened and turn into rubbles. The Japanese rose up from that wreckage and ascended with others to become, not only world powers but world superpowers. They are among those that are called by the UN as very very highly developed nations today, while countries that were at ease then never came to those heights. They turned water that was not inhabited by fishes and undrinkable to living waters for fishes. Whereas, other nations, which had many fishes in their waters when Japan had their environment destroyed by the Second World War, today have no fishes in their water.

Today, Japan is among the first tourist destination for improvement experts, because of its current sophisticated industries. They have the highly rated number one car company in the world (Toyota), and are ahead with the west in pursuing current technologies such as AI and robotics, and also going green. They have the best improvement journey experts but they copied or learnt many things from American occupation troops, both in the US and on their home soil. They had a very big problem and spent all their lives to create solutions. There are many countries that do not have complex machines assembling plants, let alone a manufacturing site (plant). These they do even to the smallest needs or challenges. They don't consider them as a problem.

Your need or big problem: pilot?

This book will align itself and give you the best, if you have a problem to solve. How can it be used to solve the smallest problem we have to the most complex in our society? When the mindset is built into our children or ourselves, a right source to identify real problems are areas where our health is threatened. No area of need that is not related to our life or health. Our current problems appear as if there are no problems as a continent. Also we need to take a deliberate approach towards positioning ourselves because of the abnormality that has become normality. However, if you have no immediate problem, then we can show you another method. This is done because most people are comfortable, others create the discomfort. This is an otherworldly approach where the timing is supernaturally controlled that can be learnt from the people of the east, and from princes, and Satan, and demons or the oppressors. You should learn by observing the things happening in your life and to others. That is when we realize that we are really in problem and we lack knowledge of many things. How dare you? Did you just mention Satan and devils? If you were afraid when you saw those sentences, then you have spiritual problems but you haven't realized it yet.

Suffering as Natural growth path

We as mankind lost our place to others, and are afraid of every wind that comes our way. To bring us back suffering as taught in the church was meant to recalibrate us but we dodge it and give our responsibility to others. Unless you are injected deliberately with some amount of challenges, you cannot understand the value of what problem or needs are. Violence that is it can change your orientation forever. I strongly encourage that you do this with experts or daysmen, and invisible doors will be opened to you.

"There is no man" is a common phrase in use in the bible because truly there are no men. This is chiefly because we don't know what Christ wrought for us by his death on the cross. When you try to make yourself a man, you may easily see the vacant lines of needs that need to be met in our society. Anyone that can fill up the space and understand these things should make one; and it is impossible to make a biblical man without a woman, according to God's standard. But that is possible in, and with ignorance.

Without much patience and endurance, how we cannot truly understand to position ourselves to manage needs or challenges. I don't want to quote newspapers and radio but if you check these media, countless problems that need our attention fill the whole place.

Government and Leadership

The people before government; there is no government without the people. The real idea of the government is to share resources from the center not mainly to provide resources. People are elected to effectively divide the national cake into meaningful uses. In many countries this cake like the ant will come not though collections efforts but on individual resources and taxes pulled together for the government to share and to make available social amenities. When the people aspire entrepreneurially to undertake sustainable venture that tackle need of their environment, then much more development is recorded. As in many civilized nations today. I am only trying to make you understand that there are plenty problems to solve if we can conscious register in our mind, that there are problems. Then we can manfully prepare ourselves and seek a rebirth because our problems require more than patriotic mindset to solve. For a rebirth, we need personal transformation to align ourselves with the tone of any problem. We need to re-dig the wells of Abraham, and with the freshwater, have a fresh view of the problems and tackle them.

After we have done this, before we can now seek, true wisdom and knowledge. You cannot come to a place of wisdom or deep knowledge without a problem solving mindset. Where there is desire to solve human (societal) problem there is love. These are the channels to getting the most of this book.

Service to God and Humanity

Any problem solving mind should primarily bare service to God and humanity; if you understand this you will never fail. In this is all your strength and help, even the most powerful strategy that is out there in the business world, and any endeavor. Avoid competition, our real problems lies between competition and comfort zone. Competition is a deceit and steals away solid knowledge. Economies that have solid knowledge have maintained stability in the value of their currencies over the years. In paths of excellence, and to move into endless realm of improvements, we must understand our problems and needs. Avoid false needs and problems. And that our problem is like a worm, if not tackled, continue to eat us up. But if we take care of it, there is also the reproduction part of it, and that makes us fruitful and multiplies.

Consider your ways

"The people come to me and ask me to ask for God's decision for their problem. Ex 18:15-16

and has made known to me now what we desired of You,
for You have made known to us the solution to the king's problem. Dan 2:23

"What's your problem with us? Why are you invading our territory?"- Judg 11:12-13.

LEARN BY OBSERVING

"My son, give me thine heart, and let thine eyes observe my ways." Prov 23:26 KJV

Yet so that you observe attentively, - Josh 22:5 Douay-Rheims

To do this, do what the scripture say you should do (trust and obey, and observe statutes), take instructions from the scripture. Now observe the result of the actions taken against the Word, and repeat from the Word to the Word, and observe from your dreams or revelations, and that can now be transfer to everything around you. Then also repeat from the result of your observations.

Do good for the sake of your spiritual eyes
If you don't feel the need to do good because of past experiences and troubles, please rethink your decision. Only when you receive good from the LORD and you do good to others that your spiritual eyes can be opened, Except your reception from the LORD is taken away by adversary. Be very sensitive, and that God did not say we should not expect, but while wait for our expectation, we must be careful and learn by observing, and to find the miraculous. In the beginning of your journey, you should not be worldly in learning by observing, because you will miss a fundamental principle, if you imitate others

very early in your learning to observe. This is what is meant in the twenty seventh verse of that same chapter by, *'and a strange woman is a narrow pit.'(Prov 23:27)*. In the Word such things not measured against God's will are referred to as strong drink, fornication etc.

Experiences

At the place of observation, when you get contentment right, angels and ministering creatures may attend. They have numberless functions and in many instances should bring the things to your notice. When you observe yourself and others that is how you can capture and recognize events that are re-occuring in your environment. And some of them are to your hurt not for man's profiting. Then nature will speak to you. And that the trees shall come not naked, or clothed before you. That is when every one of your thoughts should bear fruits. No sign or language, or speech that cannot be useful when fetched within that field where you are observing. Observe the effects of messages on your life as they come from the messages, not the messenger. Keep him or her out of what you seek. That way you are not deceived by impressions, which are truths.

The nature of the way

In the state of observation, you may be fortunate to find there treasure houses, hidden cities, aliens or yourself being an outcast a stranger where you have lived for many years. To do this more effectively, observe with focus on improving, inventing, innovating or creating a thing. Only when your soul is prepared that nature will reveal itself to you. At this state you will be more lucky if grace flows in. In a second an hundred page story can be given to you.

Even the life of a moth and a toad will be precious to you when you get it right. All of these must start with prayer even when you are passing this ability to little children. For our children, it is much easier, but for an adult in Africa, they may have to deal with many demons, and fears, and disorientation, and a huge mountain of ignorance.

From your office or shop floor or school to the church, there is no place where you cannot learn by observing. As for the creations, there is no end to what they can do. So are they created to serve. They see, declare and search facts out.

Renewal and regeneration

Thought not commonly discussed in personal life as a means of renewal, or regeneration, to learn by observing is a vital area for personal transformation. It is the chief of the ways of pathfinders. Do it when content and you will find secret things of God in your spiritual life. You will get streams of revelations. Get control of your dream life. *'For a dream cometh through the multitude of business.'(Eccl 5:3)*. Be awake or asleep with healthy mindset. I also hope a people's feet will be their endurance. Many people do it with high curiosity and pressure but this kind is done with a free mind not overly seeking. Listen and be sensitive to your environment.

Make it effective

Unless you come to words of truth or knowledge of truth, you will not do effective observation. Fear God, and learn by observing with natural truths. God's word is the foundation for you to do it successfully. Then you can learn from output, and to inputs. All these things to learn have living elements, and power. Unlike the scientific approach. All things respond to thoughts and spoken words. This is a spiritual journey. Again, it is the easiest way that you can mark manifestation and the word. By learning this way, no other contributions are needed. *'To the law and to the testimony: if they speak not according to this word, it is because there is no light in them.'(Isa 8:20)*. After this, to seek in scientific space becomes very easy.

I am not writing about one of your science methods. I know that many will think that way. Though you can apply it to scientific methods and it will work perfectly for you. And this is a natural or spiritual subject.

Be patient

When you are set to observe anything be patient and if possible do it over again and again. Now people will try to intimidate you both spiritually and physically but pray for strength and stand firm. Ignore shame or disgrace-make scorn of it. These attacks shall come from the most trusted, and from those you respect the most, and personalities held to a high esteem of higher ground in the society. Including places that may be described as holy. The whole world can be observed from one single point, from the east of the garden (of Eden). From the throne of all nations or dwelling place.

This book is written to also help you be serious, start your journey and also get up to that place. Herein the strongest institutions are fundamentally established. Learn to observe, then experience will come to you, and now can begin to bring out or fetch out meaning or give meaning to existing things. In the year two thousand and seventeen (2017), I was tailoring my skills and updating my resume against a UN job assignment application in Uganda when I heard the voice. *'The most profound source of knowledge is the garden',* But where is the place? I don't know how you will get there. And I only heard that voice behind me years ago. I kept it and remembered it always, and meditate there in. Now this revelation is meant to show you the importance of learning how to observe that you might find it. At the entrance of that word there was brightness into me, as a one who found a great spoil. I thought it through and through without understanding it but held it still. Neither did I also ignore the voice. Because I was not told how to find the garden, or given the true meaning of the parable when I heard the glorious voice but it came to me years later. Today, I am talking about it because I found it. If I am to tell you straightly, then you have missed all that this book can give you.

Borrow eyes to observe

One quality rule among some of the world's most powerful organizations when they have employed new graduates or new people to their operations is to allow them to share what they have observed within the company processes. Then the students or young graduates return outstanding best practices which are drafted into the company lesson learnt and their operation improvement implementation units.

This principle is difficult to practice in many places especially churches or among bosses, who cannot be changed. Give room for others to contribute and analyze input and outcome. If I am a men pleaser, I would not be a right person to observe correctly or do good judgement. Until you stop judging people from clothes and looks, and status, you will not come to the land of substances.

There are faithful Christians, who come to church with one cloth every Sunday not because he is poor, but because he has spent his living pursuing a goal in the invention or innovation room. He still has to come to church because he believe God is doing something in his life that is not finished, and he would want to prove it to the church and the world. Many such people hang in there, in shame and despised, and some of them are usually those that encourage others the most but have nothing to show for it. This is almost the same with those in affliction.

He is like a man who was cooking in the kitchen while his friends are also waiting in the living room but when he is done cooking, he came into his friend and found none. Now if you don't have money in the society, others will begin to also observe what you eat, wear, and even your hair style, whether cheap or costly. But if you have money, then others may accept all your abnormal hair, dressing and short comings. I think, among others that there are two mad men in the street that need to be carefully observed, one is, and the other is not truly mad. If we fail to observe the correct one carefully, we are going to send them both to the church or rehabilitation centre when one of them is actually needed in the creative space to proffer solutions to problems or challenges. In the land of the lost (an American movie) the doctor was found by the woman lying on the floor of his room (his lab), everywhere littered, unkept and dirty. And she would have called a doctor or pastor to pray for him. But she found him with the device that was already completed.

A great man, they say, is misunderstood. When you come to the place where everybody misunderstand you, and even call you mad, then it is easy for you to understand the oppressed mad person, who takes the blames, and panadol for others problem.

Consider your way

"To acquire knowledge, one must study; but to acquire wisdom, one must observe"
Marilyn Vos Savant
"My son, give me thine heart, and let thine eyes observe my ways." Prov 23:26 KJV

Yet so that you observe attentively, - Josh 22:5 Douay-Rheims
Hear therefore, O Israel, and observe to do it; that it may be well with thee, and that ye may increase mightily, as the Lord God of thy fathers hath promised thee, in the land that floweth with milk and honey.(Deut 6:3).

UNDERSTANDING AND WISDOM

And I have filled him with the spirit of God, in wisdom, and in understanding, and in knowledge, and in all manner of workmanship, Ex 31:3

Who hath put wisdom in the inward parts? or who hath given understanding to the heart?Job 38:36

And the Lord God formed man of the dust of the ground, and breathed into his nostrils the breath of life; and man became a living soul.(Gen 2:7)

Don't let anyone define understanding and wisdom for you. But try to get it yourself because you are the source, especially in a generation when almost everything is adulterated. And that breath you will know it, and understand it. The understanding and wisdom mentioned in the bible is what I seek for you. On wisdom, again, this is principal wisdom and must be obtained through patience. Is ok to start with what you understand by understanding but as you practically push through these things with experience you may be told by the men of the east that you do not have understanding. You will understand many things in practice, and that is the way. As you have learnt through doing that understanding come but that knowledge does not match principal things of wisdom and understanding.

What understanding and wisdom?

Let us define this in accordance with the subject at hand. Now understanding is the ability to know, and see the narrative and own it and allow it to work effectively for you and others. Do his commandment and depart from evil –serve him alright and get results continually.

Now wisdom is the ability to be able to hear God's word or voice, when it comes, and either to love it, appreciate it, be threatened by it or afraid of it but you keep and hold it still, and also able to brew it at Beersheba and dominate it. If you don't care about wisdom you have no understanding. Both wisdom and understanding can be measured.

And unto man he said, Behold, the fear of the Lord, that is wisdom; and to depart from evil is understanding. - Job 28:28 KJV

The fear of the Lord is the beginning of wisdom: a good understanding have all they that do his commandments: his praise endureth for ever. Ps 111:10 KJV

Be changed by wisdom

Most people are upside down, and ignorance is responsible. If you can turn, you will come to principal wisdom. Wisdom today is not as it is commonly understood today. Now wisdom in the hand of man is like an intelligent man, which performs functions for you more accurately without your effort. It is beyond just having knowledge. Wisdom is more than a dynamic knowledge. This is what the ancient possessed. Neither shall an army come near you; nor the most ancient throne of kingdoms in the land be hidden from you. Until we perish our old ways, and dwell on the ancient paths. In patience look, and live, and save the poor. If I have challenged you, now then pray with God's word. And dwell therein until you get answers. Listen and observe all around your life, even the most irrelevant. You come to the day, ever give up.

Living understanding

If you truly keep God's word and pray, it will come. I want to distinguish it from others; hence I call it living understanding. Everything in this book is a thing of wisdom and understanding that comes from practically doing things. So knowledge is required for you to improve findings and implement them. The last four or first four chapters will help you generate knowledge, and help you have a living understanding if you understand them. The kind of knowledge that will come to you will be distinct. Rivers of living water you would get. If you are true to yourself you will never be the same again. This book practically describe away and life that set you apart or distinguish you from your friends. You will understand almost all secret work that is out there in the creative industry. From the secret of music, other art works and many thing in the society. Why they succeeded. Because it should flow from your immediate circumstance toward a real solution and that can be continually improved. In that way you can be sure that you are understanding truths or something relative. Avoid second hand knowledge at this time. It is ok to compare but do your own too. If we have a little knowledge of angels spirits, we cannot know God. And will be left with a very poor knowledge of God. The spirit of God searcheth all things and would give you unique answers.

Be careful when you follow the herd

With what are you following the herd? The answers that the experts have ignored, and give you the things in front of your eyes that you never thought of or understood. God can render books useless because of incomplete or insufficient knowledge. Don't agree to what you don't understand as most people do. We say yes to things that we could not understand. This time because you are going to deploy the result yourself, you have to be more careful in other to avoid excesses. Spend quality time focusing on bible verses. The understanding will be given you. Get your mind cleared up, and sleep over problems. Anything that you learn thoroughly you are able to deliver confidently. If you receive profound knowledge, then you have come to the land of desire. That alone automatically engages you. Surely, you will never be unemployable anymore. In business or private institutions and in government the rare commodity is a profound knowledge. I have never seen, such people being irrelevant and stranded or out of business. Our economy is a knowledge economy, and it will always be so. Now the stories in the news are filled with issues bothering on technical knowhow. Thread on the narrow lines, and succeed in them. Until we remove dark clouds from the way of clarity. And if you have understanding, reproduce that which is always distinct toward solutions. The ability that provides service with ease – It would cast. Trust and lean on things of the Word, not on your own.

The secret of the LORD

The well spring that produces all fruit is the word, and that is the LORD. The secret of God is with them that understand Him. To assemble, and take counsel, and manage all things of life is in the understanding of God's word. Countless people never admit that God can be involved in any aspect of their lives. I can prove this. And this book can help you achieve the same goal. They that stumble at the word cannot do it.

'For the Lord hath poured out upon you the spirit of deep sleep, and hath closed your eyes: the prophets and your rulers, the seers hath he covered.

And the vision of all is become unto you as the words of a book that is sealed, which men deliver to one that is learned, saying, Read this, I pray thee: and he saith, I cannot; for it is sealed:

And the book is delivered to him that is not learned, saying, Read this, I pray thee: and he saith, I am not learned.

Therefore, behold, I will proceed to do a marvellous work among this people, even a marvellous work and a wonder: for the wisdom of their wise men shall perish, and the understanding of their prudent men shall be hid.(Isaiah 29:10-12,14.)

We see him

If you are not open to challenges, you will not advance to walk with the LORD. To many walking with God is an ancient faith. But we say Christ is the same yesterday, and today and forever. And preach you, too, as love knows how, by kindly words and virtuous life. We see him. We cannot give him because we do not have him. When he lives in us, then we can show him, to the world. When you lie, everybody knows when you are lying. And we lie because we do not – a lack of understanding. So do the hypocrite. They take you away. But he keep us when we find him. And we have what it takes to make that happen. None can deny his presence but you. Who hear and speak, and feel, and touch. Who hear him, and speak with him, and feel him, and touch him, and live with him- understand him. This can be taught, and show to anybody. How I go on, and read that children book for them is seen. For Christ, I can do. To shave hair, and wash you, and clean you – No one can deceive you, but to help a man in understanding Jesus, we show like shadows. Our thoughtless interests. Then can I rightly divide the world of truth. Am I worshiping idols? Why do peel the cover of matter in science? If am born the dry land, then unveil the universe. As you enter into his presence day and night that you come to light. From the smallest creatures, and all things, and everything that hath breathe. No man is able to understand.

Consider your ways

And unto man he said, Behold, the fear of the Lord, that is wisdom; and to depart from evil is understanding.(Job 28:28)

Let us hear the conclusion of the whole matter: Fear God, and keep his commandments: for this is the whole duty of man.(Eccl 12:13).

COURAGE

And what the land is, whether it be fat or lean, whether there be wood therein, or not. And be ye of good courage, and bring of the fruit of the land. Now the time was the time of the first ripe grapes. Num 13:20KJV

And he shall stir up his power and his courage against the king of the south with a great army; and the king of the south shall be stirred up to battle with a very great and mighty army; but he shall not stand: for they shall forecast devices against him. Dan 11:25KJV

How much do you know about your soul or inner life? This is your courage. They that know their God will do exploit. And the place of exploit and the good land of fruitfulness is in the south. Now we need the south to take a close shot at the west, then we can match into our true north and conquer territories. How shall we escape captivity forever? That is the light of courage, and what it is. *"Have not I commanded thee? Be strong and of a good courage; be not afraid, neither be thou dismayed"*. And in Daniel it is written. *"And they that be wise shall shine as the brightness of the firmament* "Your knowledge of God is the security of your destiny. Now consider the words firmament and security. And the word 'firm' in firmament. All of these came from knowledge of God. Your courage is determined by how much of God that is in you. So that you are able to go forward.

Overcoming the calling to discontinue a pathway

As what you know is tried then courage is returned. And you become firmer and firmer in your mind through experiential knowledge. Then you possess the courage to also enter the unknown and into the south. Only the knowledge of God that you have that will stand with you, when your courage is tried. Even the closest and most respected people in your life can deny or discourage you on your journeys. Now you are going into things the elderly, sages and wise in the society don't know. At the mention of the last sentence alone many will immediately give up. Many will make you an enemy, and you may also be talking with speeches that offend them more. Not that you are wrong. Courage takes you into paths which no one has seen. Also there is no doubt you are going afraid, it is a more scary thing when you deal with spirits that rob you of your God and knowledge. They may lie and deceive you, seal and unseal things for you if you are on the right path. Till this day knowledge of the ancient people in the word of God and myth exist. Most of those stories are not myth; they carry hidden and everlasting treasures. Courage is not the absence of fear but sometimes in going afraid. And make sure that you are going somewhere. Before you connect your adventure weigh all facts against opposing views. Trust not in a guide. If you are not walking with God's word, chances that you are doing the right thing are slim. To stop the mouth of lions nurture your conscience be driven by the word of God. Know that sometimes you would be acting like an alien. Because you will be talking about something that does not exist in the mind of people. Don't be intimidated by others experiences. The LORD is your strength that alone will keep you going. Patience and endurance will never yield result without having to manage sudden surges that come on the way. In biblical life there are many destinations your faith have not reached, and when in hope you behold the their gates afar off, many still require that you pass through more several gates which you never heard of before coming to light, like the onion bulb. Each with its mystery embedded. If your courage is not tested, many things will not manifest. In my life I have to struggle through trials daily over a period of time.

'Men's hearts failing them for fear, and for looking after those things which are coming on the earth: for the powers of heaven shall be shaken. And then shall they see the Son of man coming in a cloud with power and great glory. And when these things begin to come to pass, then look up, and lift up your heads; for your redemption draweth nigh.' (Luke 21:26-28.)

If I am not to go forward and be free. How that I deliver no more escape tales. When in two weeks or two years or more, a mystery is not ended, what they took the earth, out of the ancient wells, that covered a language, and written in the name of Jesus – what a great work of genius. And coined for anything that a covering on our feet, and shoes, courage led me to that, a cup hidden in a sand. Signs and wonders, in a treasure found. When you come, to the end of matter. Still a company of thorns, if lies are crooked, and to speak parables. Nor do they seek harm, neither an escape setup. How dare you? We won't trouble you, if you are ignorant. At the command of two letters, then clouds give rain. They are still here. Without fear, if you held by hand. It is a function of grace, and age that reveal the creator spirits. Not for the faint of heart. He is a dread by nature, wars, and very compassionate. But our Jesus a Lamb and a lion. Stand and behold resurrection. Quit yourself like man. Flee lies that you may see him. Pride and fear rules, now the scroll cannot be opened, the dread of ancient and young spirits. Open not. Come and see not. Before men, you have stood thousands of spirits, and challenging your madness. You are mad. They echo. Foolishman, they pressed on. Also, they defend and attack. They judge, accuse and question everything. Your parents are waiting! Wake up! The people wait for your salvation. Even the thorns ought not to last more than one day. They stare at the ignorance in your eyes, and said, say no more of it.

Go forward you perish, and stay in the defence you perish. If I perish on the forward, I prosper or die, and come to day. In very deed, fear is present selling her waves. I never know it, and was silent. Never was anything learnt, abiding in no go zone. Sometimes I was told to run; only when the situation was most uncertain, then fear came. Lo, snares and darts on my hand and against me. I am my own devil, even me. I will not know, until I am known. Without courage I have no other way of knowing. Take our help, it is torn in endless pieces, none hardly make sense. I am a companion of elite preachers, who are at ease. *"The troops of Tema looked, the companies of Sheba waited for them. They were confounded because they had hoped; they came thither, and were ashamed. The paths of their way are turned aside; they go to nothing, and perish'.* I am doing native work anyway. Without shame, without self-pity. I threw it away. Then a vision came, and a man from the east and said to me, **"O disturb raha"**. Over a narrow path, where many could not sleep, also for the fear of ghosts, and a sea of voices that spanned the ages up to Noah. People great and strong on your trail and be managed by the law or personal covenant with the LORD Jesus.

Consider your way

"That is at bottom the only courage that is demanded of us: to have courage for the most strange, the most singular and the most inexplicable that we may encounter." - Rilke on how great upheavals bring us closer to ourselves

'It is our faculty of fancy that fills the disquieting gaps of the unknown with the tranquilizing certitudes of myth and superstition, that points to magic and witchcraft when common sense and reason fail to unveil causality. But that selfsame faculty is also what leads us to rise above accepted facts, above the limits of the possible established by custom and convention, and reach for new summits of previously unimagined truth. Which way the coin flips depends on the degree of courage, determined by some incalculable combination of nature, culture, and

character.."– Maria Popova on Johannes Kepler's life.

And what the land is, whether it be fat or lean, whether there be wood therein, or not. And be ye of good courage, and bring of the fruit of the land. Now the time was the time of the first ripe grapes. Num 13:20 KJV

And he shall stir up his power and his courage against the king of the south with a great army; and the king of the south shall be stirred up to battle with a very great and mighty army; but he shall not stand: for they shall forecast devices against him. Dan 11:

LIVE CULTURE

You're a one-man defense system against this culture,....I'll back you up every inch of the way. Jer 1:18-19
And so Babylon, the most glorious of kingdoms, the flower of Chaldean culture, - Isa 13:19

When a people cannot measure or describe their internal culture it is not alive. Therefore the brain behind lasting institutions or organizations is not used. Billions of dollars are spent on cultures that are not founded and understood. However, if you want an institution that can survive all storms build a live culture. The bible says, and the whole earth was of one language and of one accent and mode of expression...and they settled and dwelt there in the plain that they found". This is the law for any organization that must be successful and preserved from generation to generation. One language, accent and mode of expression is for the organization to have the same internal mind. These things can be well understood spiritually as I have been given of the LORD's divine mercy. So the internal culture exists in any institution, they actually run things. To be one with it is to be aware of this interface and also manage it.
A nation or organization faces hardship for this singular reason that is when they are not in one with the internal. Is this the reason organizations like Ford looked for men to hire in the olden days? Not

all men are men; it is very difficult to find a man. Now it takes understanding of who a man is before you can claim to be one. And that to manage a culture in an organization from the external and internal is a thousand times more sophisticated than doing it from the external and lacking knowledge of the internal role. When we get an institutional culture right, then it should be able to outperform organizations that rely solely on smartness and connectivity. Also internal and external culturally managed organizations should rather provide a better in route for the effective integration of connectivity and smartness. This is a better case for ethnicity as solutions to successfully managing complex nations and large problematic democracies as Nigeria. Because that in a live culture the internal is aligned with the external.

In this economy among other knowledge workers like cultures are preferred and do more excellently in innovation initiatives. Speak no more of national development and catching up with others without recourse to live cultures in institutions. Therefore the internal working elements must be in place. And don't spend any other money on cultures without considering this aspect. The first quality prerequisite for a high performing organization is the culture.

This kind of culture is not common but can be develop like all others. For a live culture, that principle is embedded. A live culture is tailored with time, and needs. And the needs are organically managed. Where the processes are fluent, as in a language, is because they have one language (an internal) generated by the live culture. And said, 'I have been praying for Kingdom advancement' –for the church of the LORD Jesus Christ', I received this vision in the cause of this project, a confirmation generated by an internal workings of our organization, and to what I share now. Without this formation there would be no proper vision dream, mission, documents directives and instructions. And communication would be stymie and misunderstood. To keep a customer and truly satisfy them, our organizations need to be collectively built from this culture. For this very deed, organizational knowledge is also internally preserved in many from generation to generation.

Only on live cultures are we set to explore the universe, and solve challenges that confronts mankind. Now our current environment is very predatory to both institution as well as people. None can escape alone without fairing on an internal feedback system or ecosystem – a truly natural ecosystem. This unconventional approach need to be interfaced, and that it already exist and foreign (Strange) in nature, therefore we could not know our internal engine. Unless we see it, and close the gap between us, and interact with it and know it, we cannot build a live culture. In the traditional and best methods is hidden access to building a live and lasting culture. Every details of it is live and staring at us but we do not all have eyes to see it, neither understand that we have to befriend it. On the quest or journey for improvement, lies goldmines, with a live culture, and that the possibilities are endless. Even unquantifiable, and incremental meaningful changes that would be unmatched, for growing and expanding your existing institutions in innovations. And that the elements of a live culture is the life of any society. The quality of the vision is a function of its internal culture. If I make an institution live by reviving its inner life, then all of its documents and structure would be live, and not just a fanciful image or idols. Like, an oil flowing in a pipeline, would the entire processes flow – a living spring not automated.

From the minutest tasks, to the more complex ones, in a given place can an impact be effected. When a culture is not live. It becomes insensitive and to the environment. How an internal feedback that report and collate outcomes. Light and locate strongholds that is rarely noticeable by man. And the organization, or people is shielded, without a single cost. Never will it abide obsolete; all that makes it future, it shall gather. Now experience true peace, where others failed. Why the happiest group? Unlike many, birthed in pains. Therefore shall a people, change champions be made. As a man, that discovered a treasure house; everywhere he sees it but none else can. So the people renew, and gather diamonds and make suit for the culture. The conventional culture struggle, without the internal help. Because no understanding, and ignorance of truths. And that a live culture, at appointed, times issues reminder to valuable causes that we forgot and bring us back on the way. In this way that its life is maintained. Turn a blind eye, and everything is upset. I know that traditional settings have these things but no one is observing them. What to do and what works is before many eyes but none cares. A live culture is built for those who observed and cares about their soul. This culture is sensitive to the present and ancient environment, and our strength is in the old.

In his work, 'Of Africa' Wole Soyinka writes: *"These all constitute potential commodities of exchange— not as negotiable as timber, petroleum, or uranium perhaps, but nonetheless recognizable as defining the human worth of any people—and could actually contribute to the resolution of the existential dilemma of distant communities, or indeed to global survival, if only they were known about or permitted their proper valuation...... It is its humanity, the quality and valuation of its own existence, and modes of managing its environment— both physical and intangible (which includes the spiritual) — that remain the primary, incontestable assets to which any society can lay claim or offer as unique contributions to the attainments of the world."*

Consider your ways
You're a one-man defense system against this culture,.....I'll back you up every inch of the way. Jer 1:18-19
And so Babylon, the most glorious of kingdoms, the flower of Chaldean culture, - Isa 13:19 .

BE A DAY DREAMER

They provide their own inspiration and invent their own visions. - Ezek 13:3-4GNT

And they said one to another, See, here comes this dreamer and master of dreams. Gen 37:19AMP

Don't listen to the dreams that they invent, Jer 29:9TLB

This type is about responding by giving interpretation to dreams, dreaming of dreams, understanding and interpreting visions of dreams and visions, and managing visions and not only about sleeping and bringing forth dreams for interpretation. A day dreamer design or programme and align the external organization or personal life with the internal. God is responsible for visions and sure future, so is the gift he bestowed on Joseph and Daniel among others. These are they that can from dreams programme chaotic situations or states from hardship, and to avoid shipwreck. All is according to the message of the bible, and must be read with the same understanding to get the best of it. The gift of the Holy Spirit – if you truly understand the Holy Spirit you should be able to do the same.

Responding to dreams
What Joseph the dreamer did was to programme the lives of people

including nations to comply with the desired outcomes. Many people don't know that Joseph saved both the baker and the butcher. From their dream he repositioned their lives to live. That is one of the first secret thing in the scripture that I learnt. To be hanged in a tree, for the ignorant is death in Joseph and the two prisoners but for those that know their God it was a blessing for the two prisoners. Joseph saved both of them; for the other he rewrote his death sentence from his own dream, and received life. Joseph did the same thing for Egypt, he changed the course of history by aligning the invisible to the visible. This was a serious job but he did this by programming it, by the word of God. Joseph was not the only day dreamer in the bible, there are countless people from Adam to Jesus disciples including biblical characters were day dreamers. Every chapter of this book carries deep secret that when learnt and understood set you far above others or competitors. They are somewhat hidden, so that you can dig deep. I am also trying to put you through a learning curve.

The Spirit of God
Only God can help you do a true successful vision and dream. Because dreams and visions come from many sources. It is difficult to recognize which one is from God, except we are passed through suffering. Even as Daniel, Moses, Samuel Joseph, among others we should possess our soul in patience, then we are able to discern the spirit of God, whether we be rightly filled by the Holy Spirit or not. When we know this we should be able to teach it to little children. If we are able to look into our dreams and visions, in our waking up or sleeping correctly and confidently, then we stand in the place of discussing being a day visioners or day dreamers. While you remain in your business, and cannot tell what dream issues out of brain, or demons or why you dream a dream or what is a dream or vision, and where they came from, then it is difficult to help. From who?

A visioner and a dreamer
I would like to distinguish, between being a visioner and a dreamer,

and to take out the main purpose of this chapter. Now the path is opened for us to distinguish the genuine two. A vision cannot be drawn by a human being but can be done by spiritual man -angel (beast). You can write it but created or weaved by you.

A dream can be drawn or weaved by human being. Like Joseph, with the help of God man can alter the future by creating and writing down his desire by dreaming it – being a day dreamer in the physical world. Like an enterprise architect.

While in the case of vision, and God without your involvement sends you the vision or in response to your prayers, and you write down what is spoken or reveal to you from the spiritual world.

For the day dreamer, unlike vision, dreaming can be solely done according to your desire or God's desire.

But vision is chiefly God's instrument or thing, though it is often done in responses to your desire or need, and God's plan.

In all cases, to be a day dreamer, you are like a carpenter, and you write and fix the things the way you want it and by yourself.

As the typist only type down what he is given, so is the visioner. What you received can be used to architect a main or bigger frame in bits like the ant or in pursuit of greater objectives.

Now always, and the spirit speak the vision, what we receive, we also write down.

To dream a future (what many call visions today in organizations); you write it yourself and can also say it. When we are drowning in the sea of knowledge without God, this knowledge would be strange and impossible.

When you hear a vision, you heard the voice of swearing and it shall be establish, only hold onto it. Be careful with contrary swords. And learn their uses and able to manipulate them.

Unless streams of truthful and, spiritual actions are taken, you will never achieve your dreams, even if they are masterfully written.

But both of them can play supernaturally, one often comes in large chunks (dream) while the other (vision) is always in bits, and their outcomes does not depends on whether bits or large chunks.

One vision from the LORD can create or make a thousand dreams. You can or should insert visions into making your dreams. This today may be the big vision or achievable picture.

Without your help God can unleash the power of a vision, and is not always so with a dream.

In a single step, you can be catapulted into the place of your vision or destiny but it is not always so with a dream.

These kinds of things (dreams and visions) last forever and can be passed from one generation to another.

Vision are spiritually prepared without your help, and are like bullets ready to be shot out at the appointed time, but for some God has released them already. Or they are supernaturally activated as soon as you are received and many of such trigger as you process them in your mind or brain or try to understand it. Some will not be activated if you do not care and take no action, they can be there for years waiting for you. Also the reason why many fall victim to strange voices, and why you should also be careful, and take a second and look at them before you cast them away, because if you don't pay attention to God how can He use you? Hence the calls for courage in times like this. The ability to threaten the stylist in the list.

Unlike vision, dream is not always ready and fit. Take the gun, then gather the bullets and load it, make the right mark to target and so on and so on, then you shoot and missed some targets. But when it is done with God's help, it is like vision.

Let the visions be hidden or inserted in your dream then you have a big vision for your corporation. The dream is in your human capacity an overall picture of how you want the enterprise to be run. Without a personal encounter with God you can never setup a true vision for your life or organization or country. As you act the dream on your daily activities then more tailored visions will begin to come, and that will automate everything. Many churches have well masterfully written prayers tailored to their needs they have not moved or replaced those prayers by complete vision or direct word of God that is addressing the particular situation. When you read these visions, you address God very directly. Except, you have fellowship with the spirit, you cannot understand that you address Him directly, even with the bible also, especially in the Psalms. This is the beginning of a true encounter with the LORD.

Dreams and vision can be written down while awake with your eyes opened but the vision would be coming from the spiritual.

Consider your way

They provide their own inspiration and invent their own visions. - Ezek 13:3-4GNT

And they said one to another, See, here comes this dreamer and master of dreams. Gen 37:19AMP

Don't listen to the dreams that they invent, Jer 29:9TLB.

EVER LIVING FAITH

Behold, his soul which is lifted up is not upright in him:
but the just shall live by his faith. Hab 2:4
for they are a very froward generation,
children in whom is no faith. Deut 32:20

I know that you have heard many definitions of faith. Also the preachers have told you countless times to 'have faith' in Christ Jesus. And that you were also left helpless with those words. In the early days of my joining the church, I have never been as worried as to when I am asked to go home and have faith over the problem. Because I don't know what it is, and you left me. I know Him little or trust in the Lord Jesus but I have come to church to see the people of faith. To help me on to God but surprisingly when we have done all that we could with countless description of the subject without a real answer; I am now told to go and have faith. The greatest of such sermons or help came from Heb. 11:1, and I appear to understand it, every preacher seems to teach it. Still I could not get faith. I am born again like you, and believe Jesus as Lord and personal savior. Yet I don't have faith. Then I came to others who had faith but none left me with faith: After serious prayers I was told – go, 'just have faith'. I have followed multitude to church, and received miracles. I prayed and received answers as you. But I realized that I am not telling myself the truth, I lack understanding of the matter. If you can't describe what you have or doing your knowledge is very poor.

If you can pick up faith as you would do a pen or pairs of shoes, then you know it. In the place of 'have faith' maybe we should not say 'believe'. No, howbeit that it is not already confirmed. The work after faith actions is in believing, that's where the real doubting or more work is done while waiting. That way a faithless me, can reply 'help my unbelief'. Because I have been helped several times while they try to describe faith, belief is used to ease the burden of the misunderstanding - an escape route. If you can't describe it alone, then use trust, believe and faith together to describe the mysterious ghost. We even have to consult the best dictionaries in the world to define them;

What Faith is not

Things hoped for: Faith is not things hoped for

If you are 'faithing', and 'faithing', and 'faithing; it without an answer, you have no faith. If you are 'faithing' it, you have no faith. Because, someone, not God, commanded that you should hope for things or prolong or have them afar off. Then you hope without any immediate response or actions for an unknown future afar. Nothing is wrong with this action but nothing is as bad, if we are walking on the wrong path by our own assumption in ignorance. Let us walk on the right path, even, if we have to stand still believing when no response is received.

Things not seen: Faith is not "things not seen"

When we tell ourselves always to have faith after prayers, then we have indirectly informed ourselves that we don't know what we are talking about. We have treated it as not seen.

What Faith is: measure it

If ye have faith as a grain of mustard seed (Matt. 17:20)
If ye had faith as a grain of mustard seed (Luke 17:6)

If ye have faith(Matt 21:21)

Have faith in God(Mark 11:29) – that is not empty. I did add it?

Those things which he said(Mark 11:23)

Now we talk about the things first before we believe. In the fourth items to 'have faith in God' you must first have God, like the things he saith or substance and the evidence. All things were made by God, and the word was God – the things (evidence) he saith. Avoid confusion. Now let us look at Heb 11:1 again and the other passages.

Faith is a substance

Faith, you can have it

Faith many had it.

The easiest: For by it the elders obtained a good report. With what did they obtain a good report? Was it not in the name of Jesus? When Jesus said it is a grain of mustard seed, which is what it is. And what is a mustard seed? Where then is a grain of mustard seed? These are things even ancient living things – spirit 'the words that I speak unto you, they are spirit, and they are life. While you can take it and use it, "so then faith cometh by hearing and hearing by the word of God. Take the word and use it. You don't have to faith it to have the word in your hand or mouth. If you understand this and succeed, the mountains, storms, clouds, wings, rocks and waves will greet you. That is when you truly march towards being born again. We deceive ourselves in the pulpit when we don't know these things. If you want to hope for what is in front of your eyes. I don't have problem with you. But the scripture says *He putteth forth his hand upon the rock; he overturneth the mountains by the roots.' (Job 28:9).*The word of God is your hand. I have described the start of the journey, and will get you really busy. Only when you succeed in these few things, shall you come to the place Jesus is talking about, where you are very sure not only sure but see the mountains with which you have to deal – the mountains and what you use to overturn them. The place of the servants in the introductory scriptures. Where people like Noah, Abraham and the rest found favour with God. These things happen now in our life in more practical sense like in their days, even Adam, than we describe to day in many places. I would encourage you to watch Bishop David Oyedepo's Shiloh 2021 full programe –it was on Faith, and listen to preachers like Dr. Uma Umukpai, David Ibiyeomie and the redeemed Christian church general overseer when they preach, among many others. They all have unique and mysterious styles and very interesting to behold if you can see it. This is not a recommendation or ranking of men of God but the first places that came to my mind and where you can see this in display in live practice sermons. Our description depends also on our uses, thus our maturity – or old age. Man is capable of doing the same things the fathers did. Nothing has changed only our understanding is altered by us. When we turn to God we see these things. The calling of beasts into the ark is still possible today, and even the reappearing of creatures suitable to our environments. A more ghostly thing is the description of the Holy Spirit ninety nine point nine description of the Holy Spirit may not be correct. Read it in the next chapter on sorrow in the way.

Consider your ways

"I called my servant, and he gave me no answer;
I entreated him with my mouth" Job 19:16.

"I got me servants and maidens,
and had servants born in my house; so I was great,
and increases more than all that were before me in Jerusalem" Ecc 2:7,9.

SORROW IN THE WAY

I will greatly multiply thy sorrow and thy conception;
in sorrow thou shalt bring forth children; - Gen 3:16

but the Lord shall give thee there a trembling heart,
and failing of eyes, and sorrow of mind:
and shalt have none assurance of thy life:- Deut 28:65-66

Not a tragic call for sorrow or pains
In contemporary books your mind is elevated as the subject to manage this important aspect of our life. Therefore is kept away and one imagine an invisible science without anything to measure. Sorrow here is deliberate and a strategy but joy is not predictable. And that joy can come when you least expect it and not planned. But sorrow cannot be planned. One chose it to suffer not in sorrow, and in hope you may find rest. In the hand of the maker joy is found if you persevere. In a matter and manner of love – is it deployed?
And that these are very hard realities to accept but they don't have to dwell in your mind as negative weapons, rather it should be used as the only tool to your next break through, and they are indeed positive weapons, even now you can turn them to prayers. A broken heart is that which is dependent and trusting on the loving-kindness of the LORD. Now you should receive the blessings of a broken heart anytime, and whenever you are on those conditions.

Sorrow here is like a ladder. Because learning is a step by step process with punishments. Only that the suffering is like a sacrificial cost paid to conquer mountains of ignorance. If you do well, then you will get punishments. That he may punish not. And you will not be taken upward by each punishment, as you do the rung of the ladder. If you are fortunate aliens will administer this to you. This was Moses experience when he said; 'I have been an alien in a strange land' (Exod 18:3). Consider the word sorrow as many sowing in rows, and in a field or farming activities. When reading the bible sorrow takes place sometimes. Because the action itself is sorrowing in God's plan —when he be sought. Do it best with prayer. And if they or ministering creatures (angels) guide you through it by God's word, then all your fears would be gone. This is not to upraise sorrow or wish bad luck on anyone but to encourage those who are under suffering and to show where many contacted life. Not to know the truest meaning of the word suffering is the first problem, thus we dwell in suffering or sorrow. It is in listening in these conditions that we hear the words, even the deaf when they understand can hear it. As every word in the bible is pure, so is sorrow. Now what meaning did you give it? And that it will take a combination of hope, fear of God, love, courage, strength, belief, trust with a contrite heart – to make you stay upon your God

You have need of patience
Our courage, and call for strength, is to buy time to be patient, so the bricks are not formed from the feet. But many people are not patient enough to get to this point. Are they not experts in taking you up the ladder, and where are they? I know that right now they are saying to you if you go forward you will see these things. Here the prophet Isaiah writes: 'Thine eyes shall see the king in his beauty: they shall behold the land that is very far off. Thine heart shall meditate terror. Where is the scribe? where is the receiver? where is he that counted the towers? Thou shalt not see a fierce people, a people of a deeper speech than thou canst perceive; of a stammering tongue, that thou

canst not understand.'(Isa 33:17-19).You must be sometimes courageous enough to flee from others around when you are discouraged. If you can't go away ignore them, and do it with faith and prayer. There the LORD will see your salvation. Be strong and very courageous. Where there is no pain, no joy. No one can understand; anything here completely without putting the things to practice. Or healing of a very painful experience. The punishments are meant to stretch you, and expand your heart. Then the lights enter you, as you journey up the rungs of the ladder. Further, pains and resetting are automatically done on you without your knowledge or you commanding it. You just have to do what you have to do. Only that we are exposed to the reality by it. Still the right path is usually not filled with many challenges especially when you choose it yourself. But endure and be patient.

"Only the soul that knows the mighty grief can know the mighty rapture. Sorrows come to stretch out spaces in the heart for joy". Edwin Markham.
"A hand fall means a high bounce – If you are made of the right material" – Anonymous.

These quotes are facts. Never were they written to please your reading appetite. But are things to encounter if you are on the right journey. He will keep in perfect peace whose mind stayed on thee. Shall the inhabitant of our hidden village walk with us? He is faithful who has promised, the LORD Jesus is true to his word. But shall we truly be patient, so that he may be found to hold us by the hand? How then shall we see the place of wisdom or value it, and even know it. In this type of sorrow which is a journey, as we may do with every other sorrow, until we are led here; let us live in the truth of it, so that we are not burdened by the meaning our wrongs understanding of the word gave us and move to its truest meaning.

Daily sorrow and joy

None can turn sorrow by day into joy. Until a measure of sorrow is consumed, joy will not come forth. Unlike joy which you have in hope, from there you may even come to abundance of it. Stay upon your God in your challenges or pains, and you may progress to where it is now deliberate, and deployed. Some from mighty grief, and not to mighty rapture. Because of the strength and abilities that comes at the end. Many, they have no knowledge, they could not understand.

Wickedness and Load

But I have seen sorrow in the place of enjoyment, and it was confused to be absence of joy. There is extra wickedness that may come in place of joy. As a man travels hundred miles instead of one in ignorance without knowing, so is that extra load of wickedness in the place of righteousness. In Ecclesiastes Solomon writes: *'And moreover I saw under the sun the place of judgment, that wickedness was there; and the place of righteousness, that iniquity was there'.*(Eccl 3:16) This is not the type of sorrow I am talking about. It is a deliberate evil. Then you have to do a secret fervent prayer. And that it happen in the place of bringing forth. Not sure but could be because of misunderstanding and differences. And when you are also carrying load and others afflicting you, while they dwell in ignorance and blinded. Now a deliberate sorrow is faceless and formless, controlled and managed by the saints. By such means were countless people justified. All that is written here is pure in the eyes of the pure, and the sorrow is taken away. This is true: *Unto the pure all things are pure: but unto them that are defiled and unbelieving is nothing pure; but even their mind and conscience is defiled.'*(Titus 1:15). Howbeit, walk in the direction of the Holy Spirit. Be careful. That I heard in several occasions from the saints whenever others' interest is never encroaching, and then fear was on every side.

Who is the Holy Spirit? Who is the Holy Ghost?

If you can see with your two eyes then you can see the Holy Spirit. Don't allow anyone to send you out of church because you don't have the spirit – a lie, by what even measures by speaking in tongues, visions, laying hands will only turn to a great lie. False measures Are you having problem describing those? And that a five child old can describe. If they cannot describe it, they don't know it. It is not nuclear science. Be careful. Trust not in a guide. It's indeed a narrow way. Isaiah writes: *'For my thoughts are not your thoughts, neither are your ways my ways, saith the Lord.'* (Isa 55:8). It is a very scary thing, even a consuming fire that burned into the lowest hell. There is no respect of persons. Where is the place of true gold, and sapphire, and diamond and the rare treasures? And secret truth, unseal that make you supernaturally play therein. Therefore a lost world suddenly appears before us. Sit still. By the quality of what you hear or see shall you come out victorious and refreshed? Be a deliberate listener to the environment. Don't wait until you see problem or suffering before you listen to the world. Put your ears down, and you will hear the world, like the psalmist; my ears has thou opened in affliction. Learn to listen rather than cry, and hear the world.

Like open wound, sorrow call up our reality, and keep us naked with God's will and plan, where our real money and economy can be found. That is why, some say, in my life sorrow has become joy and that death has become life, and that life is not hard. If you ever live in this truth, then you can easily come to the state of blessedness. When we are separated by unforeseen or foreseen circumstances into unwanted recess or wanted recess, let us make maximum use of it. Let us not dread that precious moment of our life. I answered you. Have you found the differences yet? And I know you heard him, by what can you describe him? But I was not writing in the Holy Spirit. No, not a book on Holy Ghost. Heard him still, sit still. Now it appears like wickedness, so shall thy light enter you. Don't just sit there. Moses writes: *But if from thence thou shalt seek the Lord thy God, thou shalt find him, if thou seek him with all thy heart and with all thy soul.* (Deut4:29)I answered you. Were you not angry? No sorrow? If you were not angry. Is ok. This is in letters, and when the work is finished with you in the letters and spiritual there is also practice where suffering may take turn on your finances and business.

Consider your way

AFRICA IS THAT STORY THAT WE TELL OURSELVES

I will greatly multiply thy sorrow and thy conception;
in sorrow thou shalt bring forth children; - Gen 3:16

but the Lord shall give thee there a trembling heart,
and failing of eyes, and sorrow of mind:
and shalt have none assurance of thy life:- Deut 28:65-66

As the days wherein the Jews rested from their enemies,
and the month which was turned unto them from sorrow to joy,
and from mourning into a good day:- Est9:22

Sorrow is better than laughter:
for by the sadness of the countenance the heart is made better.- Eccl 7:3
"People wish to be settled; only as far as they are unsettled is there any hope for them." –
Rapdo Emerson on our resistance to change and the key to true personal growth.

JOY IN THE WAY

So that the people could not discern the noise of the shout of joy from the noise of the weeping of the people: for the people shouted with a loud shout, and the noise was heard afar off. - Ezra 3:13 KJV

When the morning stars sang together, and all the sons of God shouted for joy? Job 38:7

Touch Joy physically

Joy comes forth after weeping. Where did you hear noise of joy? And was the noise heard afar off? Howbeit, you have gone to church or at home and with others shouting but you have never heard the noise of the shout of joy that you invested so much time in. Now a state of being happy, and laugh, and smiling is not a reception of joy, and that happiness is not joy. Desire it, and I say, pray that you hear it. You can touch joy. In the path to living a meaningful life, it is the joy you receive on the way, which makes the differences. When I am affrighted by a frightening wind that mountains raise and trouble and overcome new spears and swords. I also knew there was overcoming of it. At the end of each phase, the joy that comes with it surpasses the ongoing rising and falling experience. If I do not encounter a wicked problem, I may not know the value of celebrating at successful end. And even may miss the strength to undertake the next assignment that comes from past success. But they will be in pains and disappointments who did not understand, especially the closest people to you. As you go further apart, you will be in opposite direction to many. Moreover, your days and nights are calculated by the director. Until the morning break forth you remain in night. And that indeed joy comes in the morning. You should receive many more mornings, and joys, and more to understand. Job writes: 'If they obey and serve him, they shall spend their days in prosperity, and their years in pleasures.'(36:11). Why joy? For many, these things are known or given to you after intense suffering. I know that some inherit joy from parents.

How I know joy

It will not come easily like that, what are these challenges that I did not mention above? I was under a spiritual surveillance unknown to me in a long time. At the peak of this surveillance which has lasted for three years. By the peak, I refer to these three years. Overall we are looking at twenty years captivity that a secret corporate culture did more than monitor me. And in these years, the spirits operated as if they were allotted every hour of the day in twenty four hours to perform one thing or the other. And they make you interested, and disinterred. And that they were determining to run everything in my life contrary. Also, I would declare that I share these things in the integrity of my heart. I understand joy, and that my ears were opened in this period of affliction, but I have to battle and endure the voices, as for the voices, some for mercy, some for correction some for lies, some for weapons against me, some for weapon against my enemies, some for deceit, some for delays, some for blessings and prayers, some for wisdom, some for instructions, some for prophesies, some for reporting, some for signs and wonders, some for accusing others and some for exposing all the secret and work of the power of darkness, some for praises, some for money, some for healing. None can write and identify what each of the voices bare without getting lost in the letter wonder of each, whether it is for evil or good. They were well written according to biblical standard, and revealing the word as it is written. Well written, whether it is for good or evil. By the power of the wings they can be caused to go East, South, West and North.

Look and live: guided with eyes fixed on Jesus
Whatsoever each new challenge comes a joy, a fitting joy to match it or terminate it. That is why we must always be with the bible and that there is no other place to find it. When you do things, you suffer and when you don't follow the next instruction, then you missed the light. It is in the light that countless lights are also revealed. This is a circle like the sun and moon. And that seasoned changes throughout climatic period infer. Still many circles are before the child of God, even circle of failure and success. Our prayers and fellowship are so built. When you learn by observing the minutest details of what you do in faith. Until you overcome little challenges, and have a change of attitude toward yourself. Never to expect any encouragement where you are hurt the most or in its dark seasons? If your house is flooded during a flood, no amount of sun shine, neither heat can restore-dry it, in that same period. Then as a dream of the night, is passed a circle with thanksgiving and praise, and with joy before follows. Never is a rain without a flood first. The floods caused a land to be sown before the rain in a barren or desolate land, or where there has been no water.

And tormentors abuse these things, to add or increase the sufferings and delay the time. They control you by the wind but in all these things in use, we are more than conqueror. For we overcome daily, by the blood of the lamb, and his mercy toward us. A people great and many that were capable of many things. I ran to the church but the people did not believe me. They draw rain and ride on countless chariots.

They keep me from my sleep, even in the night. And make my heart hot with fire every night till the morning. They count the number of spoon of rice and the eba, I swallow, if it exceed a particular limit, then I am purged. In a miserable affliction with prayers, I was so joyful when I found where they are. I am for peace but they are for war. For the mercy of the LORD I was able to learn many things about them. I know that some will ask; while in silence when you face these things? As for the people closest to me none of them believe me. They were with me day and night, even now as I write this book. Many throw darts, some prayers and their intentions are well concealed in the use of the language. They are very crafty to the point of frustrating the writing of the work also. They hide around to talk with you. And they play games with people's lives. They monitor farms and businesses that they might perpetuate evil with both hands. We cry and pray day and night but they ignore the cry of the children of God and still continue to lie against their neighbours. They are very good at lieing against neighbours. When they say that they were sent by a particular neighbor, they I say I will go and talk with the neighbor – they disappear or stop using that neighbor immediately.

Consider your ways

The blessing of him that was ready to perish came upon me: and I caused the widow's heart to sing for joy. Job 29:13

"Therefore the redeemed of the Lord shall return, and come with singing unto Zion; and everlasting joy shall be upon their head: they shall obtain gladness and joy; and sorrow and mourning shall flee away."(Isa 51:11

BE FOCUSED

In conclusion, brothers, focus your thoughts on what is true, noble, righteous, pure, lovable or admirable, on some virtue or on something praiseworthy. Phil 4:8-9CJB
Don't focus on the stubbornness of this people, or on their wickedness or on their sin. Deut 9:27-28 CJB

Make your eye single
When you work with truths from single live path, also strategically associate and stay on them to pursue a better state. To observe, learn and understand the power of focus do only one thing chiefly well - Here is a hidden secret along ethnic lines because of adaptation. And one of the most powerful tools in national development is therein. Unless you focus your energies, you will never be a good observer. Also you can attract supernatural help, when you do one thing well. Listen to your listeners. These things I am saying to you are things, and if you want to find out, meditate on what you are now reading, again and again. Do one thing and direct your energies toward improving it. You will never work again in your life if you get it right. Because you will enjoy whatever you are channeling it to. Many people who have confirmed to this truth on focus said it is their secret of success. Pastors, Engineers, Doctors, teachers among other claimed this fact. They were made and became pacesetters.

In the place of bringing forth

To know God or learn more about Him be very focus, and your life will not remain the same again. This is the easiest way to raise your head above the noisy world. Now these actions will make your life. When embarking on this journey set your mind on truths. And gather them together, and they can only come to you when you take steps in their direction. To those who do, not those who only hear. As you take steps with kin observation they will come to you. Truth is like a door when you find one open it, and in it, you come to another, again and again. Truths are difficult to find, for the mind that realize and cares to find it. Now you hear about truth since you were born but you may have not heard it. On the day that it opened to your eyes, and then you actually start to truly live. The fastest place to locate them is in the bible. You can be a Christian for a thousand years without knowing what truth is. When heard the hearer bears fruit accordingly. If you are not in truths you are not focusing. How then shall truths come to wisdom and understanding? Don't be ashamed, if you are not sure of knowledge of truth seek out somebody who knows and ask, or pray to God. All the principles that run every organization or mastermind behind many things have underlying truths. If you know them you will not be afraid when they appear in different forms. Show me, any successful work, and I will tell you the secret behind it. For us to be acquainted with truths, we must make some effort to understand our inner life. Focusing requires the whole of your being to come to a state of profound meditation.

But do it in truths

Go to church and serve God with all your strength, heart, mind and soul. Do what Christ said you should do. And observe, again and again and pray for it. You will come to the knowledge of truth. These biblical truths are the foundation of anything out there in the world. Therefore give heed to them that make trouble with their eyes and in truth. Consider an enemy that openly and persistently distract when focusing. Some set free from the yoke of a pen when you set your face as a flint against them. As a man weep, so is that which produce little. None give it, and have many genuine ways. How do you get too much afraid? Never be choked on seven grounds, not even in eight. Gather truth, many of it, and use them to build a rail. Without it, you would be left out, and distracted. Consider focus as a train on a rail line –if it is not this. It is nothing else. At the end of every journey, then the terminal reveal a new track of rails. As you do this continually, know that countless people will observe it. A natural way that work all things; If you stay in it. Only when the process reaches an impasse or end (the point of tiredness), still the new rail will emerge from the former. In movement, and number, and time, and sound, and sight, and taste, and smell, and touch, and feeling, and among others, which you can use to collect truths. Avoid help that will delay you, or keep you out of track. No focus team should succeed without a rail of truths. Now light is what sustain everything, and give the life of the whole work. Where precisely is faith? That embraces grace, and the purest form of our faith.

Distraction and challenges

When the rain is not failing on focus projects, remain calm and wait patiently, and endure any deliberate wickedness that is set to set you off balance. And this prepares you, so that you are able to withstand a tougher distractions and challenges. Specifically the days that is mercifully granted you to stand before the LORD of Spirits. Consider this as mercy whether it be deliberate or naturally from Adamic sin. For those who are in falsities or evil, it is not so, and recognize it not. Again, learn it deeply, and put your trust in the LORD. If you meet the right people, then you get to know more. It is through affliction because it is so commanded that one easily meet the Lord Jesus. But many cannot come to this place; If Joesph was not sold into Egypt he wouldn't have had all those blessings attached to him. Only the few who will see the mountains, the raging seas, and the roaring lions. But you may obtain favour, when you in these conditions seek the Lord, sit solitary. The Ishmaelite merchant men's business is always contrary to us, so that we may be saved. They sit at the gates, and many reach where they are someday if they remain steadfast.

Don't be deceived

There are many deceivers who lead people and are in very high positions whose role is to work with truths but without an encounter with it. When you are on the right path, and walk in truth and on the way. Don't be deceived, and seek personally the things of the LORD Jesus. God can help you with godless words. The type that will worry you. For me, I have experience it many times. I know that everybody dream dreams, even terrible and fearful revelations. Be focused and prayerful and very vigilant. These things change as you grow. I have seen answered prayers hidden within the usual fearful dreams. If you are distracted by family and friends, you will never get to see these amazing gifts. Until your few days will come, be focused and prayerful. But listen to those on the opposite view sometimes, and retain important feedbacks.'(Phil 4:8-9).

Consider your way

In conclusion, brothers, focus your thoughts on what is true, noble, righteous, pure, lovable or admirable, on some virtue or on something praiseworthy. Phil 4:8-9CJB
Don't focus on the stubbornness of this people, or on their wickedness or on their sin. Deut 9:27-28 CJB

"This is my Son, marked by my love, focus of all my delight." -2 Peter 1:17
The wicked hope to destroy me,
but I focus on your instruction.
I see the limits of all perfection, Ps 119:95-96CJB.

MEASURING

And ye shall measure from without the city on the east side two thousand cubits, and on the south side two thousand cubits, and on the west side two thousand cubits, and on the north side two thousand cubits and the city shall be in the midst: this shall be to them the suburbs of the cities.(Num 35:5)

Numbering
By making judgment against not, only God's acceptable standard, but also by the accuracy of its measure. Unless you measure by God's word your knowledge of measurement is poor. The Egyptian economy was measured by Joseph at a period that has less technology like ours. He did it at a very young age and was able to manage the whole land in line with God's principle of numbering and leadership. 'And Joseph gathered corn as the sand of the sea, very much, until he left numbering; for it was without number.(Gen 41:49).He preserved and saved a nation from destruction with simple numbering. That is what the bible said when you read it in the literal sense. After this in his leadership style and management, how it was done, is clearer in the spiritual sense? As a Hebrew, he practice and deploy his dream programming skills which he learnt from the prison house. Only when you live from the internal sense of the word, then

you will be exposed to the secrets that were used. Of this, by God's mercy, Joseph observed and was helped, and the dream worked supernaturally. By measuring anything can be taught excellently. Today, our current civilization is delivered through mastery of measuring science. In this book it is written to draw you from a sea of confusion and to let you know it that all started with God. And a child of God, will have access to such an extravagant grace that we don't know is available in His word. Now you cannot please God always without understanding it. All our works are in measure. Only that for some it was all gathered already, and was released in great measures.

Creative Work and Measuring

But wisdom in measurement is the same in all creative work. And that one nation is better than another in its ability to measure accurately, and in what they can measure. Whatever I can measure, I can create it. And to create is to collect, and the ability to restore or refine it, and to bring it nearer or simpler to its natural state, and work with it. If we can do this, then we can make many things out of that which is created. Many things in their purest state can automatically perform unique functions that we have not known. This is the law of all inventions. From the bible I have the principle of the tree which helps you to collect and measure; the principal of AI in which you are helped to understand all its properties and make use of that which is created, and the principle of water and Fish on which to manage its survival, and improvements. In AI, if I can completely measure anything about it, then I can fully describe it. And that will give you an edge to manage it and its interactions, and to be more efficient, and effective. Also these principles can serve in any venture. By understanding measuring we can advance towards creating, making and forming things.

Can everything be measured?

Everything including love can be measured. By measuring we can make everything we do on daily basis count. We can set our time against the work, and business and things that we don't have. Then we can minutely set ourselves on meaningful ventures. Also quantitatively learn new skills or with craft slowly create things, and raise an empire. The difference is in the measuring and setting personal targets. But we must be keenly committed to our plans. By measuring, replicative abilities are secured. Then we can work on speed and agility or intelligence.

'All qualities can be expressed quantitatively, 'qualitative' does not mean unmeasurable' –Gilb Tom

'In physical science the first essential step in the direction of learning any subject is to find principles of numerical reckoning and practically methods for measuring some quality connected with it.

I often say that when you can measure what you are speaking about, and express it in numbers; you know something about it;

But when you cannot measure it, when you cannot express it in numbers, your knowledge is of a meager and unsatisfactory kind;

It may be the beginning of knowledge, but you have scarcely in thought advanced to the state of science, whatever the matter may be' – Lord Kelvin 1893

'Everything should be made as simple as possible, but not simpler' – Albert Einstein

False hopes

There is no worse mistake in public leadership than to hold out false hopes soon to be swept away – Winston S. Churchill

I have spent the past seven years collecting spiritual facts and truths and principles on my career. Never fail to confront the brutal facts. Even in a man's life, I discovered that our knowledge of spiritual things must be narrowed down to truths from obedience. At this point that religion wields great creative wonders, and contribution to societal development.

Today, I was motivated to keep weaving this book by other assets standing dormant and wasting away. I thought to write it in bits that add value to the life of the readers. To equal the value of this book to that asset; By measuring what I do and that the reader also can do. Again, I am creating a tool, whose value is right now more valuable to me than any asset I have ever seen. When I consider the value myself, and that the asset can bring to the table. And that the actual meaning of a single rich ancient art can achieve more in a year than many physical assets. Now the secret of one word in many places should cost more than millions of dollars. Not even the secrets of notable cultural arts in many ethnic regions of Nigeria, which are highly valuable but not known. And I can do the same in what I termed other 'alien punctuation', and with secrets hidden in unfamiliar and familiar folklore, and films, both local and foreign. The truth is that they can't be valued for money. I can use these things to teach you the difference between the Old Testament and the New Testament and money making. Why then are people afraid? These are free things of the LORD Jesus Christ. They are before your face and not demonic. Not selling them and you will find others who excelled in kingdom work as you see of John Milton, John Buyan among others in the world, see the next chapter. In our current literature, science and any other field today, if you mention the best name in Nigeria and another place in the world I will show you the secret of the work or success from the bible. For literature I can do that with Soyinka.

Consider your ways

Then thy elders and thy judges shall come forth,
and they shall measure unto the cities which are round about him that is slain: (Deut 21:2)

But thou shalt have a perfect and just weight,
a perfect and just measure shalt thou have:
that thy days may be lengthened in the land which the Lord thy God giveth thee.
For all that do such things, and all that do unrighteously,
are an abomination unto the Lord thy God.(Deut 25:15-16) KJV

AFRICA IS THAT STORY THAT WE TELL OURSELVES

And with many such parables spake he the word unto them,
as they were able to hear it.
But without a parable spake he not unto them:
and when they were alone,
he expounded all things to his disciples. Mark 4:33-34 KJV
The measure thereof is longer than the earth,
and broader than the sea.
If he cut off, and shut up,
or gather together,
then who can hinder him?
For he knoweth vain men:
He seeth wickedness also;
will he not then consider it? Job 11:9-11 KJV

Therefore hell hath enlarged herself,
and opened her mouth without measure:
and their glory, and their multitude,
and their pomp, and he that rejoiceth,
shall descend into it. Isa 5:14

.

LOVE AND CHARITY

I love them that love me; and those that seek me early shall find me. - Prov 8:17

but by love serve one another. - Gal 5:13

Be generous: Invest in acts of charity. Charity yields high returns.- Eccl 11:1

What is Love?

Love is the cause of fear, a genuine truth. In the light of it, without charity you would have no love, and your understanding of love will be very poor. God is love. Love is a spirit. When you love him, be truly committed to it. Don't abadone your name of Jesus. If you are not committed to living a life of love as you understand it, you will not come to charity. Which form is as describe here by Scott. Of love, this is one of the best descriptions outside the bible that I have found among many books that I have read. In the Road Less Travelled M. Scott writes: "But what is this force that pushes us as individuals and as a whole species to grow against the natural resistance of our own lethargy? We have already labeled it. It is love. Love was defined as 'the will to extend ones's self for the purpose of nurturing one's own or another's spiritual growth" When we grow, it is because we are working at it, and we are working at it because we love ourselves. It is through love that we elevate ourselves. And it is through our love for others that we assist others to elevate themselves. Love, the extension of self, is the very act of evolution. It is evolution in progress. The evolutionary force, present in all of life, manifests itself in mankind as human love. Among humanity love is the miraculous force that defies the natural law of entropy."

Stretch yourself against entropy in that direction because it is a hidden part often not observed by many. This is a typical example of narrative, and knowing it, and owning the narrative. Here in Scott's work, Love though wonderfully described. We still fall short when we don't know what love is in the word and owning it. If we know it then we can own it. Now try to know it and own it, if you know it you can measure it. If you can't measure it you don't know it. You need no further description or analogy because the meaning is fully within this chapter. Love is a thing of swearing or vow and being committed to it. And making sure it works. For Isaac, is said of Rebekah, 'and he loved her' (Gen 24:67), also for Jacob his son is written, 'and Jacob loved Rachel'(Gen 29:18). Herein is the secret hidden? You too can love, and stretch yourself toward another. The bible did not say that they loved each other but Jacob loved Rachel. To keep an agreement or contract between two companies, we both need to stretch ourselves toward the opposite. Not as liking where one can take interests in a thing. But in love the two people are partners under common interest to exchange and make it work. The string of love is not made perfect by a space of waiting or making to heal, and commitment and faithfulness. In case of couples, love should be built up and maintained by continuously affirming and showing such commitments or delivering health with the word of God.

Practice Charity with the help of the Holy Spirit

Charity is more of an inner life, resulting from obeying God's word. Trusting in God completely is like jumping off the cliff without seeing a help on the way down. What a great belief? You will never fail but you will find that leads to charitable life. Unless this happen to you, to understand that charity covers multitude of sins may be very difficult for you. To deal with multitude of sins is to also get acquainted with our spiritual life and the cleansing and reconciliation that happen there. Because of charity, we even receive countless gifts of love from that life. Every act of charity is done within a time frame, and charity is done best as ordered by the LORD. Not according to the way of the Old Testament law but by faith. And that such work is gathered and substantiated. Every great work out there that succeeds is because of charity and love. According to various researches and google search the bible has the highest number of printed copies more than any other book ever publish for this reason. Because that the bible, conveys true workable knowledge of God, and His loving-kindness, both to the rich and the poor equally -No respect of persons and graciously without partiality.

Give even when in straits

To get all you want make sure that your ways (services) pleases the LORD and mankind at the same time. Now the access to charitable work may be the suffering you experience when trying to do it. That is a good currency in exchange and the salt. Anybody can love and do charity with this mindset. If you are poor, and hungry, please give. Everyone doesn't use his condition as an excuse. It was in this state that many discovered these secrets. They were not rich when they stretched out their hands. If you will not give because you are poor, then you may remain poor forever. This singular truth can usher you into a realm of endless truths and favors. Many when they found out hid it but could not use it as a means to uncover, an strive for others. Open your eyes and look through that window. It is a sea of ancient pathways to your inheritance. I know that you will try it with expectation but the presumptuous adventurer will casts his bread on many waters. And that the journey is more exciting than little expectations. Therefore take you and give yourself to the leading of the word itself, and will never be distracted by little expectations. Who are you? That is what you seek. When you discover who you are; you will be ashamed for the idols that you had followed, and expected to have. From where you are now, start casting your bread and ask God for direction if you want to be formless and more effective with it. It was in the course of charity work that many learnt how to receive answers to prayers. When you do good, you do charity. The place where you can easily do good, that is so commanded in the scripture is though charitable work. By revelations on where or who to help. Be specific to state that you want to do charity in your prayer life. There are few things that we can do in the bible, here we can say we are doing good, charity is one of them. This is not attached to love. The problem is to know what is charity. Avoid debate and foolish arguments on the topic of love and charity. To do this, simply pray before embarking on it.

Honorable steps: benefits into free graces

Every commandment of God is with a blessing. Another example of such a command where good is mentioned is marriage. The scripture says 'he who finds a wife finds a good thing. And that is why marriage is said to be honorable. Now honorable means you are automatically entitled to a blessing not empty societal status without mercies of tangible good from God. If you have not seen or received anything regarding this promise in your marriage life, then go to God in prayers, surely you will find packages of favors laid up for you. You will receive physical blessings or your desire. Many couples don't know this secret. You may decide to devout time to lay claims to it when you pray. I also know that woman according to the scripture is the love of good but the honour or blessing in marriage is an entitlement of the couple. Below are two example of honorable persons or those that were due for a blessing:

Now Naaman, captain of the host of the king of Syria, was a great man with his master, and honourable, because by him the LORD had given deliverance unto Syria: he was also a mighty man in valour, but he was a leper.(2 Kings 5:1)
Arise, get thee to Zarephath, which belongeth to Zidon, and dwell there: behold, I have commanded a widow woman there to sustain thee.(1 Kings 17:9)

'But unto none of them was Elias sent, save unto Sarepta, a city of Sidon, unto a woman that was a widow. And many lepers were in Israel in the time of Eliseus the prophet; and none of them was cleansed, saving Naaman the Syrian.' (Luke 4:26-27)

Charitable people like Naaman and the Serapta woman are people who are due for blessing, because they are carriers or have good readily deposited in them, of favor by virtue of that charitable work. For Naaman it was not a title or his status in Sysria. This is not the case with love. Charity is without such agreement but where commitment exists. Here it is voluntary, with deliberate punishments coming to the doer, but who has committed to bear them. In love the punishments may not be intentional or deliberate. Punishments come out of our weaknesses and external influence on our life. When love is overstretched it may also yield charitable results. That is why people say that charity is love in action. Charity is measured by a different kind of measurement. What is specifically the measure of charity? Because charity is stretched toward the spiritual or invisible world, and that it is also stretched from the spiritual to the physical is described in the story of the Good Samaritan, while love is done in the physical world or the visible. The love of the LORD or God is not stretched to the spiritual world in the name of Love, and, it is done in the visible or physical. When you have done charity, then the LORD can also express his love from the invisible more often. Now charity is periodic but love is not. The Love of the LORD is also the love of his word, thus love of God who is in the invisible. Now you can't do love without stretching your hand toward another. Also you cannot do charity without stretching your hand toward the recipient. These things you can analyze from their scriptural uses. I have chosen to write about charity and love because of the blessings that come if you do them. Paul described charity in the epistle to the Corinthians, another proof that charity is not the same as love. He writes:

and yet shew I unto you a more excellent way. (1 Cor 12:31)

*'And now abideth faith, hope, charity, these three;
but the greatest of these is charity'.(1 Cor 13:13)*

*'Follow after charity, and desire spiritual gifts,
but rather that ye may prophesy'.(1 Cor 14:1)*

*And the scribe said unto him, Well, Master,
thou hast said the truth: for there is one God;
and there is none other but he:
And to love him with all the heart,
and with all the understanding,
and with all the soul, and with all the strength,
and to love his neighbor as himself,
is more than all whole burnt offerings and sacrifices.(Mark 12:32-33)*

Further Insight

I know that charity is not love. And that spiritual things gathered together when you do charity. But follow love that which you know. When light comes, you should understand the difference. If I am awaken by love, and then there is the same thing. But I found love teaching pupils. Then I spent my life helping them to get through their weakness without realizing it. And now I can measure it after twenty years. As I conclude this chapter, because Jesus and the father lives in us, when you are in love and present in a place, God will be present, and then God's love will also be present.
And we have known and believed the love that God hath to us. God is love; and he that dwelleth in love dwelleth in God, and God in him. 1 John 4:16

Consider your way

For God is not unrighteous to forget your work and labour of love, which ye have shewed toward his name, in that ye have ministered to the saints, and do minister.-Heb 6:10

Therefore, your majesty, please take my advice: break with your sins by replacing them with acts of charity, and break with your crimes by showing mercy to the poor; this may extend the time of your prosperity.' - Dan 4:24CJB

And Jonathan caused David to swear again by his love for him, for Jonathan loved him as he loved his own life. 1 Sam 20:17 AMP

The wings of the ostrich wave proudly, [but] are they the pinions and plumage of love? Job 39:13 AMP.

AN ENCOUNTER WITH THE WORLD

Who hath given him a charge over the earth? or who hath disposed the whole world? - Job 34:13
and he hath set the world upon them.-1 Sam 2:8

All chapters: Practice Sermons
This book require reading and practicing, and I again, I hope a people's feet will be their endurance. If you have never acted on the instruction or obeyed God's command in course of reading the bible along with this book, the gates of the world will not be opened to you, hence there is no way you can make proper sense of this chapter. In Transformative Experience, L. A. Paul writes: **"Many of [life's] big decisions involve choices to have experiences that teach us things we cannot know about from any other source but the experience itself."** This is a strange world and if you have read from the first chapter to this point without a little different understanding of the book and world, then you should begin again and again, until you get it. You cannot know and own the narrative without knowledge of these things. When you enter into faith, you will learn more about the word of God. Then you will have real encounter with the Lord. Because everybody talk about encounter with the word without touching its nature. I am talking about

exercising faith or battling with the word, and be led beyond answered prayers. And that you discover the world where things as written in scripture are seen. Take with you faith (things) from the scripture with which to do this. And that faith is not a state of your mind or to be in the spirit. But faith is the objects or pictures in your mind, colour(words) or express things of the mind. Again, I repeat faith is a substance not unseen things. If you desire that stand in prayers, wash you, and you will find these things. Don't be deceived. All the myth about the Holy Spirit will be chartered by the time you finish.

The Kingdom of God and your life

But I want to write about how the world will look like when you succeed or if you are privilege to learn about our world through commitment to Christ. In Mathew Jesus taught us saying, 'But seek ye first the kingdom of God, and his righteousness; and all these things shall be added unto you.'(Matt 6:33). The scripture promised a permanent Kingdom of God, not a kingdom to find and set aside after finding it, then return back to it later. It's going to be your knew life forever. Here your life is built and sown in Christ, and all other about it will now grow and run from there. On this the Psalmist writes: 'Who laid the foundations of the earth, that it should not be removed for ever'. (Ps 104:5). The kingdom of God is your foundation. How will you stay away from those who tell you to set heaven aside for worldly work and business first? Now, you journey conformably to the new life in Christ. Therefore you have a life of God and his ways to continually guide, lead, guard and help you prosper and journey through life successfully, and then beholding endless wonders and mysteries. No, heaven set you up, and help you live on earth, while it is guaranteeing your heavenly candidateship. At this point of fellowship is our daily living solution, where you know that science is at the mercy of the bible, thus easy with children of God. The land of uses where creating, making and forming is the order of the day.

What are these things?

I know that you will know these things if you come here. In the hymn Ancient and Modern revised 168 writes:

> *Two worlds are ours*
> *The mystic heaven and earth within,*
> *Plain as the sea and sky*
> *There is a book, who runs may read*
> *Are pages in that book to show*
> *Which heavenly truth imparts,*
> *How God himself is found*
> *Within us and around*
>
> *The saints, like stars, around his seat.*
> *But where it lights, the favored place*
> *By richest fruits is known*
> *The everlasting sea proclaimed*
> *Echoing angelic songs*
> *The raging fire, the roaring wind,*
> *The Spirit's viewless way.*

You will know these things, if you come here. In a revelation of Jordan, I met a man having a conversation talking with a very small creature, but when I tried to talk to the creature it flew round and round me distractingly without responding back to me. Then I probed into the matter and asked the man why she refused to talk with me. And the old man said, if she is there, she must be a witch: Because I knew the creature was there but refused to give me an audience. These creatures in that world are everywhere in the physical world and also dwell among us. Man must reawake to seek his lost glory. It is much more beyond Sunday miracles. They occupy two worlds, and one is our lost home. The hymn writers above have also attested to this. I have also discovered farms or fields as written in the bible. Isaiah lamented about fields, when he said, "Of a truth many houses shall be desolate, even great and fair, without inhabitant: yea, ten acres of vineyard shall yield one bath, and the seed of an homer shall yield an ephah". There are many farmers daughters, and you might be lucky if you find the right one. In fleeing away from Satan and demons, we have left ourselves and old homes behind. Satan is as big or bad as the description you give to him. Chariots and strange movements are common but not noticed between the physical and spiritual world. The children of the east are also having a tour of both worlds. I often say that the affluence and elegance surrounding the personalities around the pulpit do deceive many of the people being pulled out of the pit. The moment cannot rightly describe the suffering of the preacher or his journey and the poor. The Arabian shows that our knowledge of God's ministering spirits is wrong, that's if you even care to know. Do we really care to know? Or we have now created a lie in our head, and we describe the lie as the real thing. And we see them far off than near. Like Paul we could say to them I know where you are, and even to any little child the truth can be clearly shown them. The battle with Satan is fierce because we don't know the angel. How can you know Jesus if he appears again without notice and you don't know the angles?

But now mine eye seeth thee.
Can you confidently say that also to your generation? The chief reason that many will never know Jesus or encounter him is fear. What if his form we cannot know? And wickedness in high places. Where is the high places? When we say far above principalities, could it be afar off or as far as your thought could carry you? Or what you make of it. And the dwelling place of the wicked can be found? Where now is your hope? Specially, that place of the righteous. In the book of Deuteronomy and numbers is mentioned a strange people that the children of Israel met on the journey. The giants from Bashan, Moab, Heshbon, Edom – the Eminms, Anakims, Horims, Zam-zamims, Avims among other.

When the children of Israel was sent to spy out the land of Canaan, it was the Anakims which they saw, and some of the spies were scared to death and brought back an evil report. And they said, 'and we were in our own sight as grasshoppers, and so we were in their sight'. What a heart breaking report? Is this the story we also tell about Satan, Lucifer, devil and demons in the spiritual world? I would also like to describe Og king of Bashan, it is said that his bedstead was a bedstead of iron; nine cubits was the length thereof, and four cubit the breadth of it, after the cubit of a man.' As it is said here of their wives, and children and much cattle; 'But your wives, and your little ones and your cattle. These describe actual occupation, and of things relating to our daily living today. In another place, 'for I know that ye have much cattle'. And I decided to show people that are so described. Where is their dwelling in the world? They represent arcana of heaven and things that exist today. I am only trying to say that everything about the people described is about the LORD Jesus, and if you agree with me that we described Jesus, then the things described there exist today, exactly or accurately but in parable were they spoken in the scripture, and as they are described. When you have an encounter with the world through the knowledge of our LORD Jesus, you will see these people. They are here. If I doubt the existence of these things as they are then I also doubt where it is written in Hebrew, ***'Jesus Christ the same yesterday, and today, and for ever.'(Heb 13:8)***

THE BOOK: OTHER EXPEREINCES

And no man in heaven, nor in earth, neither under the earth, was able to open the book, neither to look thereon.
And I wept much, because no man was found worthy to open and to read the book, neither to look thereon. Rev 5:3-4 KJV

The book

Now the current church is very challenged, because of the lack of understanding to the meaning of these things and fear. And where the problem with fear is not the fear but not knowing what fear is – not a biblical language deficit. Against the bible, according to many non-Christian or pagan or business writers, there is no book existing today that has better business and war strategies that is ever written than the Word of God (Bible). Now, shamefully, we also accept such statements written in some of the most powerful business books in the world by vain beliefs. This we do shamefully because we don't know, and we just say amen. But by these very statements the writers are very confident of practice, and that the bible can be so used to manage institutions or organizations in the world. And that it is truly more than any of the best ever written from ancient China, and any other oldest classics. How is this possible? And to prove and show you is this book written.

All that is the word, which, the world is; the heaven and the earth within that we talk about daily. The four head river flowing, masterfully woven by the maker in four-phase and practical operations as described in the mysterious and everlasting Garden of Eden. And if such a thing truly exist, then the beast, cattle, fishes, creeping things, birds, fowls, wing thing, seas, trees, metals and people, all that, both in actual and in the parabolic meaning are present today. The singular reason that all nations are still hung on the garden today. And what is God's plan for man? This is the message of the LORD Jesus Christ, and his parables. The hidden kingdom, the heavens, hell etc are real.

Of an encounter with the world in Spurgeon's Morning & Evening, Charles Spurgeon writes:

"my most pleasant days are shadowed into nights; and the flood-tides of my bliss subside into ebbs of sorrow; but there, everything is immortal; the harp abides unrusted, the crown unwithered, the eye undimmed, the voice unfaltering, the heart unwavering, and the immortal being is wholly absorbed in infinite delight. Happy day! happy! when mortality shall be swallowed up of life, and the Eternal Sabbath shall begin.
"Escape For Thy Life; Look Not Behind Thee,
Neither Stay Thou All The Plain;
Escape To The Mountain, Lest Thou Be Consumed."

Works of Flavius Josephus: The Wars of the Jews Or The History Of The Destruction Of Jerusalem:

"and I have written it down for the sake of those that love truth, but not for those that please themselves [with fictitious relations]. Of which history, how good the style is, must be left to the determination of the readers; but as for its agreement with the facts, I shall not scruple to say, and that boldly, that truth hath been what I have alone aimed at through its entire composition."
for they were in a rage at the injury that had been offered them by their exclusion out of the city; and when they thought the zealots had been strong, but saw nothing of theirs to support them, they were in doubt about the matter, and many of them repented that they had come thither. But the shame that would attend them in case they returned without doing any thing at all, so far overcame that their repentance, that they lay all night before the wall, though in a very bad encampment; for there

broke out a prodigious storm in the night, with the utmost violence, and very strong winds, with the largest showers of rain, with continued lightnings, terrible thunderings, and amazing concussions and bellowings of the earth, that was in an earthquake. These things were a manifest indication that some destruction was coming upon men, when the system of the world was put into this disorder; and any one would guess that these wonders foreshowed some grand calamities that were coming."

Of an encounter with the world in his works the Holy War John Buyan writes:

"In Universe, and can this story tell.
Count me not, then, with them that, to amaze
The people, set them on the stars to gaze,
Insinuating with much confidence,
That each of them is now the residence
Of some brave creatures: yea, a world they will
Have in each star, though it be past their skill
To make it manifest to any man,
That reason hath, or tell his fingers can.
But I have too long held thee in the porch,
And kept thee from the sunshine with a torch.
Well, now go forward, step within the door,
And there behold five hundred times much more
Of all sorts of such inward rarities
As please the mind will, and will feed the eyes
With those, which, if a Christian, thou wilt see
Not small, but things of greatest moment be.
Nor do thou go to work without my key
(In mysteries men soon do lose their way);
And also turn it right, if thou wouldst know
My riddle, and wouldst with my heifer plough:
It lies there in the window. Fare thee well,
My next may be to ring thy passing-bell."

AFRICA IS THAT STORY THAT WE TELL OURSELVES

Of an encounter with the world from his works Paradise Regained, John Milton writes:

I, WHO erewhile the happy Garden sung
By one man's disobedience lost, now sing
Recovered Paradise to all mankind,
By one man's firm obedience fully tried
Through all temptation, and the Tempter foiled
In all his wiles, defeated and repulsed,
And Eden raised in the waste Wilderness.
Thou Spirit, who led'st this glorious Eremite
Into the desert, his victorious field
Against the spiritual foe, and brought'st him thence
By proof the undoubted Son of God, inspire,
As thou art wont, my prompted song, else mute,
And bear through highth or depth of Nature's bounds,
With prosperous wing full summed, to tell of deeds
Above heroic, though in secret done,
And unrecorded left through many an age:
Worthy to have not remained so long unsung.
Now had the great Proclaimer, with a voice
More awful than the sound of trumpet, cried.
Repentance, and Heaven's kingdom nigh at hand
To all baptized. To his great baptism flocked
With awe the regions round, and with them came
From Nazareth the son of Joseph deemed
To the flood Jordan--came as then obscure,
Unmarked, unknown. But him the Baptist soon
Descried, divinely warned, and witness bore
As to his worthier, and would have resigned
To him his heavenly office. Nor was long
His witness unconfirmed: on him baptized
Heaven opened, and in likeness of a Dove
The Spirit descended, while the Father's voice

AFRICA IS THAT STORY THAT WE TELL OURSELVES

From Heaven pronounced him his beloved Son.
A Saviour, art come down to reinstall;
Where they shall dwell secure, when time shall be,
Of tempter and temptation without fear.
But thou, Infernal Serpent! shalt not long
Rule in the clouds. Like an autumnal star,
Or lightning, thou shalt fall from Heaven, trod down
Under his feet. For proof, ere this thou feel'st
Thy wound (yet not thy last and deadliest wound)
By this repulse received, and hold'st in Hell
No triumph; in all her gates Abaddon rues
Thy bold attempt. Hereafter learn with awe
To dread the Son of God. He, all unarmed,
Shall chase thee, with the terror of his voice,
From thy demoniac holds, possession foul--
Thee and thy legions; yelling they shall fly,
And beg to hide them in a herd of swine,
Lest he command them down into the Deep,
Bound, and to torment sent before their time.
Hail, Son of the Most High, heir of both Worlds,
Queller of Satan! On thy glorious work
Now enter, and begin to save Mankind."
Thus they the Son of God, our Saviour meek,
Sung victor, and, from heavenly feast refreshed,
Brought on his way with joy. He, unobserved,
Home to his mother's house private returned.

Of these things and experience Bishop David Oyedepo of the Living Faith Church, mentioned in his message "unveiling the treasures in a revival", when he said,
"For the joy that was set before him.
The church shall become the global centre in the world.
That is the light at the end of the tuned.
We are in that season now.

AFRICA IS THAT STORY THAT WE TELL OURSELVES

That is where we are going. If you see it, it will impact on the quality of your engagement.
The thinnest unbeliever that chose to be dedicated because an envy of his generation.
You can't see this and need to be pushed.
My prayer is that you see this thing today that you become partner with Christ.
Those who knew you before will be wondering whether you are still such a person.
They supernaturally empower to operate at the realm of Christ
A season of supernatural favour
A season of restoration of health and wholeness.
A season of financial portion Revive thy work O God.
A season of supernatural change of level
"The sole umpire of your due point is God"
They stayed on, stay on.
My share of challenges: You are never made a champion without some number of punches.
When I am due, I will not be denied, until I am due I am not due.
Champions of strange order will be rising this week.
Many global stars will rise up form our midst
My soul desire is that I see you touch what I never touch.
When God swears upon you the struggle is over."

In conclusion

This book prepares you to discover the way or the place by yourself. Only the true way to the place: or source of true novelty or inventions. So that we can all join those who create and make the world a paradise.

For the youth or all aspirants in Africa, who are you? We are all priests or prophets?

'But ye are a chosen generation, a royal priesthood, an holy nation, a peculiar people; that ye should shew forth the praises of him who hath called you out of darkness into his marvellous light;(1 Peter 2:9)'And hath made us kings and priests unto God

and his Father; to him be glory and dominion for ever and ever. Amen.'(Rev 1:6)

Again, this book is a guide and a code written partly in blue. *'And shall put thereon the covering of badgers' skins, and shall spread over it a cloth wholly of blue, and shall put in the staves thereof.'(Num 4:6)*

Robert Collier in the secret of the Ages writes: *"If you happen to know some middle aged couple who are looking forward hopelessly to a dreary old age, who live in constant dread of loosing their livelihood, show them the article. There are thousands of abandoned farms scattered all over this great country of our –abandoned because the young folks want the light and life of the great cities. Many of them can be bought for a song, or rented on shares. To the men and women who are willing to endure present hardship for future peace and comfort, they offer opportunity. They offer home."*

For example, this Collier's text can be responded to with a psalm and a hymn and not a careful tilling of, as a farm, even as a field. Be dedicated to endure and seeking the face of the LORD and you would understand these things as part of the mysteries of the word of God. Now the point I try to make is that these things written by Robert Collier was written in parable. This may be something you are doing already but you don't even know. If you take it as a farm and spent time to cultivate and pay attention to it you will receive response from the LORD. I want you to mature to know it. Like the hidden Kingdom of God said by Jesus in the bible. Have you not read this in the scripture?
'Again, the kingdom of heaven is like unto treasure hid in a field; the which when a man hath found, he hideth, and for joy thereof goeth and selleth all that he hath, and buyeth that field.

Again, the kingdom of heaven is like unto a merchant man, seeking goodly pearls:
Who, when he had found one pearl of great price, went and sold all that he had, and bought it.
Again, the kingdom of heaven is like unto a net, that was cast into the sea, and gathered of every kind: (Matt 13:44-47)

That's why these works of Paul, Jude, James, Milton, Buyan, Josephus and all the saints including those who wrote the hymns were hidden. When you come in you will be able to write hymns too and easily. How that almost all those who found the way, and neither hid it? It is so commanded and those things are written in blue. Even that many of the things written, here in this book.

Africans Genius in heaven
There are saints in heaven, Africans who did not know Christ who are very wise. Who went through observing the traditional not evil part because the traditional is the express descriptions of the law? How can you know Christ without the law? Your traditionalist hold vital meaning to things of the word in true worship of God in your community if you know your God you can obtain these things without getting involved. Still, Christ is the answer there; he was slain from the foundation of the world. Many countries celebrate such things they preserved today. The problem with those who stand on the grace without knowing these things is that because of their confidence also fall short and are chiefly those referred to as people with PRIDE. Love and do charity to all with a good knowledge of the word of God. At the end our faith and the grace we received establish the law. Paul writes:
Do we then make void the law through faith? God forbid: yea, we establish the law.(Rom 3:31).

Because of that pride that they don't even know the Jesus they believe even when they have power and have seen miracles. The reason people are not writing Hymns today. These things I am talking about are clearly expressed by the hymns without this knowledge you can only write an outright soulless book. You receive signs and wonders singing hymns because they belong to angels and many were written in Jesus name. If you read the Book of foxes to see how the heathen priest killed the disciples you will understand, why is it that they were all kill by heathen priests? Hand, head and mouth, beasts, cattle, fowl, feet etc are not only as you understand them because they were passed down from the ancient. The interpretations are the same from Genesis to revelation. These things also are in the strange world language and that the language is so coined with the word of God and with power and meaning. If I have a library that cost me a billion dollar to set up, and I possessed no thorough understanding of any language and its tongue, then it is all a waste. But If I have a solid book, wherewith I can understand, and know the tongue that is worth more than a thousand of such costly libraries. Whether second or first language of English or Spanish, French and minor as in Nigeria Urhobo, Isoko, Ibibio, Ikwere, or major as in Yoruba, Igbo, Hausa, even Ashanti in Ghana, please make sure you understand your language and own it.

To paraphrase the quote of the sage Sufi Aba Said ib Abi-I-Khair more than nine hundred years ago:

"Until college and minaret have crumbled. This holy work of ours will not be done. Until faith becomes rejection, and rejection becomes belief. There will be no true Muslim or Christian – we will never come to the true knowledge of our God."

This calls for a change in that our sure and perfectly built mindset, which has also affected and made our view of the world. I don't think it has anything with killing of people. If you diligently follow and obey God's word. Then take this book with you as a companion along with your bible. Our little time is in our darkest night: the end in the plagues.

If you want to get the most out of this book make a commitment to forgive wrongs, yea, even wrongs darker than death; Your worst foes could be your best friends. Who is the enemy of the church?

In the integrity of my heart have I written these things, and do not wish to arrange a proper book that would put you in suit. Every disorderliness in this book that you have observed is intentionally in scripted or done, if according to you any exist in it.

Consider your ways

Who hath given him a charge over the earth? or who hath disposed the whole world? - Job 34:13
and he hath set the world upon them.-1 Sam 2:8

The beast that thou sawest was, and is not; and shall ascend out of the bottomless pit, and go into perdition: and they that dwell on the earth shall wonder, whose names were not written in the book of life from the foundation of the world, when they behold the beast that was and is not, and yet is.-Rev 17:8

APPENDIX: THE TRANSLATIONS OF THE HOLY BIBLE USED

A&M -Ancient and Modern revised Copyright © 1922

AMP-AMPLIFIED BIBLE Copyright © 1954, 1958, 1962, 1964, 1965, 1987 by The Lockman Foundation,

CEV-Contemporary English Version ® Copyright © 1995 American Bible Society.

CJB - Complete Jewish Bible Copyright © 1998 by David H. Stern

Douay-Rheims - Holy Bible: Douay-Rheims Translation PC Study Bible formatted electronic database Copyright © 2006 by Biblesoft, Inc.

GNT - Good News Translation - Second Edition Today's English Version Copyright © 1992 American Bible Society.

KJV - The King James Version Electronic Database. Copyright © 1988-2006, by Biblesoft, Inc.

NLT- New Living Translation ®, copyright © 1996, 2004 by Tyndale Charitable Trust. Used by permission of Tyndale House Publishers.

THE MESSAGE: The Bible in Contemporary Language Copyright © 2002 by Eugene H. Peterson.

TLB - The Living Bible Copyright © 1971. Used by permission of Tyndale House Publishers, Inc.

ABOUT THE AUTHOR

Francis Otolo spent a secret recess among black monks in the darkest parts of Africa. They held him by his hand, as they pass on these mountains, well into no conclusion and over the period.
From the saints, he learnt how to save people and businesses threatened by the two most dangerous enemies of any society.
In 2016, the consultant, teacher, preacher when he lost his job, he dropped the idea of a second hand God and decided to personally seek God, and was found of him. On life's difficulties, he received light. When, he devoted himself to the study of God's word.

With seventeen years of experience in managing for world-class quality in organizations, Francis weighed current universal working principles against God's word, and discovered that the only true ingenuity in everything that exists from God's word.
And if anyone is found of him, he can understand the source of all hidden mysteries in all spheres of life and in any tongue and nation.

Francis was a member of the research team at OES that participated on the 'first global assessment on the current state of organizational excellence' that has been supported by the OETC, GBN and ISO/TC 176. And as a manager of Quality/Organizational excellence, he has helped many organizations from different industries to leverage unique ideas, principles, tips, tricks, basic concepts, models, and methods that practically work and that have stood the test of time, to make improvements, and generate more revenue, and stay relevant, and achieve strategic goals; and product quality, and process/operational excellence.
He designed and implemented Integrated Management systems (IMS) to meet ISO 9001, OHSAS 18001 & ISO 14001(in organizations of different sizes), for clients such as Total, Shell, and Mobil which operates the OIMS (Operation Integrity Management System /Safety Health & Environment) requirements.
Even using quality management/ lean tools, he currently runs experiment on the African farmed catfish for farmers. He also consults for schools and raising their performance, and in moving toward the path of excellence.

www.ingramcontent.com/pod-product-compliance
Lightning Source LLC
Chambersburg PA
CBHW070241220526
45465CB00004B/1471